*More advance* 

# From Oversight to Overkill

"If you're interested in clinical research, read this book! It is not a diatribe against IRBs. It's a call to action to make them more humane. It is highly readable and persuasive."—*Rhode Island Medical Journal*

"*From Oversight to Overkill* is quite simply a 'must have' reference for libraries reaching general-interest patrons and health professionals. Ideally, it won't just repose on library shelves, but will be recommended to reader and healthcare discussion groups for its many insights and lead to lively debates among healthcare professionals and users of the American medical system."—*Midwest Book Review*

"Simon Whitney has studied the workings of IRBs for decades. In *From Oversight to Overkill,* he offers eye-opening stories and careful analysis showing how this well-intended system has gone astray—an urgent message for health care professionals, scientists, policy makers, and ordinary citizens alike. The book is very even-handed, long overdue—and superb."—John Mueller, Professor Emeritus, University of Calgary

"*From Oversight to Overkill* is both informative and quite engaging, making a basic case for IRB reform that is thought-provoking and persuasive. It deserves to be widely read and discussed."—Kim Eagle, Albion Walter Hewlett Professor of Internal Medicine and Director of the Cardiovascular Center, University of Michigan Health System

"*From Oversight to Overkill* should be a must-read for regulators, review boards, and investigators who aim to balance patient safety and autonomy with the need to promote responsible clinical research to improve patient care. The book is authoritative yet readable, and the recommendations provide a basis for much-needed constructive discussions about how best to reform the cumbersome oversight and review processes that can be barriers to modern clinical research."— Thomas R. Martin MD, Emeritus Professor of Medicine, University of Washington

"As a pediatrician and senior clinical researcher, I recognize both the value of IRB structures and the way they create disproportionate burdens and failed opportunities. *From Oversight to Overkill* provides a wealth of examples to illustrate these benefits lost as well as a construct to improve this dysfunction by recognizing the key value of societal benefit."—Karel Allegaert, KU Leuven, Belgium and Erasmus University, Rotterdam, the Netherlands

"*From Oversight to Overkill* focuses on biomedical research ethics regulation in the USA, but it should be read more widely, since the American model with all its pathologies has been imposed on its international research collaborators. Whitney's call for a better balance between the protection of research participants and the interests of research beneficiaries should resonate with scientists and regulators around the globe."—Robert Dingwall, Emeritus professor of sociology, Nottingham Trent University, UK

"When medical research is delayed or blocked, we all suffer. As Simon Whitney shows in this insightful and readable book, the potential for good from research is currently impaired by a lopsided system. Thankfully, *From Oversight to Overkill* also offers a remedy for this unacceptable imbalance."—Paul Glasziou, Professor & Director, Insitute for Evidence Based Healthcare, Bond University, Australia

"IRBs were created to protect patients, but they often no longer function in that way and sometimes seem focused more on protecting institutions than patients. In *From Oversight to Overkill*, Dr. Whitney provides deep insight into this hugely important subject, illuminating a system that often goes awry."—Razelle Kurzrock, MD, Professor of Medicine, Medical College of Wisconsin Cancer Center

"While their creation was initially well-meaning, some IRBs have become so burdensome that many investigators are turning away from human exploration to animal research, where navigating the regulatory hurdles is easier. Simon Whitney has done a great service in bringing attention to needed IRB reform."—Mark J. Rice, MD, Professor, Department of Anesthesiology, Vanderbilt University Medical Center

"The best book I have ever read on the problems plaguing medical research today—a comprehensive and fascinating look at how the current regime hurts patients, and what to do about it."—Scott Alexander, psychiatrist and author of *Astral Codex Ten*

# From Oversight to Overkill

*Inside the Broken System That Blocks*
*Medical Breakthroughs—*
*And How We Can Fix It*

## Simon N. Whitney

 Rivertowns
BOOKS

Printed in the United States of America • April 2023 • I

Hardcover edition ISBN-13: 978-1-953943-21-7
Paperback edition ISBN-13: 978-1-953943-22-4
Ebook edition ISBN-13: 978-1-953943-23-1

LCCN Imprint Name: Rivertowns Books
Library of Congress Control Number: 2023900174

Rivertowns Books are available from all bookshops, other stores that
carry books, and online retailers.  Visit our website at
www.rivertownsbooks.com. Orders and other correspondence may
be addressed to:

　　Rivertowns Books
　　240 Locust Lane
　　Irvington NY 10533
　　Email: info@rivertownsbooks.com

To Patty

# Contents

# Timeline of IRB History

**1947:** Judges in trial of Nazi war criminals issue The Nuremburg Code, first effort to define ethical principles for medical research.

**1953:** James Shannon of NIH creates the Clinical Research Committee (CRC), the forerunner of today's Institutional Research Boards (IRBs).

**1959:** Henry K. Beecher publishes "Experimentation in Man," proposing voluntary peer review of research.

**1966:** In Public Health Service Police #129, the federal government mandates IRBs in research institutions nationwide.

**1971:** IRB rules are modified to eliminate review of IRB decisions by university officials.

**1972:** The horrific abuses of the Tuskegee Study of Untreated Syphilis (1932-1972) are publicly exposed.

**1978:** *The Belmont Report* sets forth ethical principles supposed to govern medical research.

**1998:** Gary Ellis, head of the federal Office for Protection from Research Risks (OPRR), is grilled by hostile members of Congress. Ellis responds by shutting down all research at Johns Hopkins.

**1998–2001:** "Death penalty" era of extreme punishment for institutions that violate strict IRB mandates.

**2000–2008:** In a series of cases, the Office for Human Research Protections (OHRP) uses its authority over IRBs to halt or delay research studies.

**2017–2018:** Driven by years of complaints from researchers and physicians about IRB overreach, OHRP makes modest— and inadequate—modifications to its regulatory system.

# 1

# Unintended Consequences

DOCTORS DIDN'T DO A LOT OF EXPLAINING IN 1963, but Nathan Fink, who was a patient in the Jewish Chronic Disease Hospital in Brooklyn, was still puzzled when two doctors he had never met appeared at his bedside. The older one, Chester Southam, held a syringe and said he was going to do a skin test. He didn't ask permission. He simply rubbed an alcohol swab on Fink's thigh, slipped the needle in, and injected some fluid just under the skin.

Soon Fink noticed a tender nodule at the injection site. The doctors visited regularly over the next several weeks to measure the nodule, which grew to about two and one-half inches, then shrank. On their last visit, no trace remained, and Southam congratulated Fink on his "good resistance." Resistance to what, he did not say. Months later, a newspaper reported that twenty-two patients in the hospital had been experimented on without their consent. Fink read the article and realized that he had shown good resistance to an injection of live cancer cells.

Those cancer cells had a story of their own. In 1951, doctors at Johns Hopkins were treating a Black woman named Henrietta Lacks for cervical cancer. Without asking for permission—or even informing her of what he was doing—Howard Jones, her gynecologist, sent a portion of her biopsy to George Gey's lab. Gey was a Hopkins scientist who had been trying for years to keep human cells alive outside the body, and with Lacks's cancer cells he at last succeeded. These cells were to become one of the essential materials of biomedical research. He called them HeLa, after Henrietta Lacks.

The scientific establishment gave no further thought to the woman from whom that tissue was taken or to the family that lost her to cancer. Journalist Rebecca Skloot, in *The Immortal Life of Henrietta Lacks*, uses a rich archive of photos and anecdotes to bring the woman to life and show the shabby treatment her family received during the decades in which her cells were transforming medical research. It is true that she received good medical care, and her treatment as a source of research material was not unusual for the time: A doctor of that era who planned to give a snippet of tumor to a colleague would see no reason to ask permission or even mention it to the patient. Still, Lacks's story raises important questions about the protection of privacy, the distribution of the fruits of research, and how medical care is, or is not, provided to the poor.

Scientists had long abused minorities for medical training and research. The most infamous experiment was conducted by the Public Health Service in Tuskegee, Alabama, where PHS doctors diagnosed hundreds of Black men with syphilis, then spent forty years withholding any treatment, simply watching as the men sickened and died. The eventual exposure of this barbaric experiment created a crisis in the nascent field of research ethics, which took as a major goal the prevention of unethical research.

Today, research subjects are protected by committees, located at each research institution, that are generally known as Institutional Review Boards. When the research is meant to develop new drugs, medical devices, and biologics, IRBs operate under the supervision of the Food and Drug Administration. This book is about the far less well-known system run by the federal Office for Human Research Protections. IRBs operating under the supervision of OHRP oversee all federally-funded research—work that might involve drugs that are already FDA-approved, or work that uses no drugs at all. For more on these two systems, see the source note at pages 215-216.

In the United States and abroad, research oversight addressed the urgent problem of protecting research subjects. People who enroll in research today enjoy robust protection from experimental abuse.

But there is such a thing as too much protection, and the IRB system has created a second crisis. While IRBs work hard to safeguard subjects, these boards seem to have forgotten that all of us, including the people who participate in research, also want the better medical treatments that only research can bring. Scientists who want to use ethically sound methods to find better ways to treat patients with heart attacks or who need ventilator care (to mention just two examples) face a gauntlet of IRB demands that too often delays or damages their work.

Research oversight done right protects you from being injected with cancer cells without your consent. Oversight done wrong— review that is more restrictive than needed—reduces available treatments when you are wheeled into the emergency room with chest pain and a wracking cough. If your lungs fail, the doctor who puts you on a ventilator may follow the most up-to-date guidelines, but if research oversight delayed their development, in a sense you will be getting yesterday's care. If the ventilator does not save your life because an IRB slowed development of a useful improvement, a measure of responsibility falls to the IRB system.

Although the difficulty with IRB oversight is virtually unknown to the public, it is a massive problem. Every year, IRBs and their federal supervisors—the most powerful influence on medical care that most Americans have never heard of—delay and damage billions of dollars of medical research.

The problem stems from an elementary ethical error. Research ethics focuses on the undisputed fundamental principle that it is wrong to harm others, including research subjects. But there is a second principle that the field is largely blind to: the ethical imperative to *help* others, which includes doing research that relieves suffering and saves lives. It is wrong to sacrifice the few for the sake of the many, which can happen when the rights of research subjects are ignored. But it is just as wrong to dismiss the welfare of the many as unimportant. To put it more concretely, there is moral value in protecting a premature child from blindness or an old person from Covid. As sociologist Carol Heimer writes, "The biggest ethical lapse in American regulation of human subjects' research is the death and suffering that has resulted from slowing the pace and altering the focus of research and squandering research funds."

Yet the harm caused by restrictive oversight can be hard to see. In late 2022, with American Covid deaths passing one million, it is impossible to identify the invisible hundreds or thousands who would have survived if better treatments were available. In contrast, Nathan Fink, Henrietta Lacks, and the men of the Tuskegee study were identifiable individuals who could be victimized because there was no system of research oversight. Now we have such a system, and it claims its own victims—patients who die when regulation slows science excessively—but the names of these people are lost, mixed in with those whose death from illness or injury could not have been prevented. No doctor can name specific patients who would have survived if ventilator care were better; no journalist can show their photos and tell the stories of their shortened lives.

In response to the first crisis, ethics demanded that we see experimental victims as worthy of our care. This second crisis calls on us to expand our compassion to those who die because IRBs block life-saving research. The scope of the problem is reflected in calculations that suggest that the delay of a single heart attack trial in the 1980s resulted in thousands of unnecessary deaths (a case we'll review later in this book).

Even apart from the IRB system, biomedical research is highly regulated. Scientists must comply with rules that specify how they can access federal funds, how a study proposal may be modified, how experimental animals must be cared for, how to guard against scientific misconduct and conflict of interest, how to ensure workplace safety, and much more. Many of these regulations, however necessary, have serious deficiencies when they are put into practice. But while the IRB system is not the only burden that scientists bear, its benefits are the most hypothetical, its costs the most exorbitant, and its rationalizations the most bizarre.

In this book, I will review with you the history of the creation of the IRB system, its descent into dysfunction, and the fascinating dance between the officials who run the system and scientists who want to save lives. Next I document how IRB review damages and delays research in specific areas—including kidney stones, heart attacks, and the care of premature babies—while remembering that it affects every area of medical research. Finally, I explore the reforms that will open the door to life-saving research while still protecting subjects. I want to lift the curtain on an invisible and unknown process that profoundly affects the care you receive when you fill a prescription, are rolled into an operating room, or are flown by helicopter to a regional trauma center.

I want to emphasize that research oversight was a necessary response to a real crisis, and that we still need and benefit from IRBs' concern for individual research subjects. Ethicists of the 1960s and 1970s responded to an important moral challenge; I applaud their deep concern for the frail human who is helpless against the power of the state. Skloot begins her book with the warning of Holocaust survivor Elie Wiesel, "We must not see *any* person as an abstraction." Chester Southam violated this prohibition, candidly admitting that he hid the truth of the cancer cell injections from Nathan Fink and the other subjects in his study because he was afraid they might refuse to participate.

Henry Beecher, a Harvard anesthesiologist, thought that Fink had every right to refuse. Beecher was also angry when he read about researchers who had deliberately infected institutionalized children with hepatitis. The scientists claimed that every child was bound to get the disease, which was endemic in the institution, and so the experiment did them no harm. Beecher dismissed this argument, and in 1966 he published an article attacking the hepatitis study, Southam's injections, twenty other experiments he considered unethical, and a scientific culture heedless of the harm it inflicted. His exposé, which showed the public that science was not a costless march from one breakthrough to the next, marked the beginning of the modern era in research ethics.

Southam's cancer research was a black eye for its funder, the National Institutes of Health. James Shannon, director of the NIH, knew how easily scientists can fool themselves about the ethical hazards of their experiments, and in the 1950s he established a panel, the Clinical Research Committee, to review proposed studies at the NIH's own hospital. The CRC protected subjects' welfare and rights: It blocked experiments that were too dangerous, and it ensured that people were given the information they needed to decide whether or not to enroll in a study. If Southam had worked at the NIH's hospital in Bethesda, and if the CRC had permitted his research, it would have required him to enroll only subjects who had been fully informed and given a chance to decline. But Southam's employer was Sloan Kettering Cancer Institute in New York, which required no such thing.

Before this debacle, federal officials had made no attempt to extend CRC oversight beyond the walls of the NIH's own hospital, deferring instead to investigators' judgment and any oversight, however feeble, provided by their institutions. The Southam scandal forced Shannon to break with this tradition, and in 1966 a federal directive required institutions receiving federal funds to establish committees to conduct local oversight—the committees that we now call Institutional Review Boards. They can be found in hospitals, universities, and medical schools from Honolulu to Miami, and any scientist who wants to conduct research involving human subjects must obtain committee approval first.

Because IRBs have sole authority to reject proposed studies that they consider unethical, they are a choke point for almost every significant medical advance. In itself, this is not a problem—it is exactly what Shannon intended. What Shannon did not anticipate— in fact, what he very much hoped to avoid—was the exaggerated level of protection and the diminished regard for public welfare that are IRB hallmarks today. When scientists write about their IRB problem, they often describe the system's unreasonable delays and unnecessary expenses; yet time and money are not the system's primary costs. The biggest costs are paid by you whenever you become a patient, since IRB decisions limit the tools doctors have to treat you.

Ethicists, scientists, and regulators have spent years in fruitless argument over the IRB crisis. All recognize that the system has serious problems, but none can agree on the explanation for these problems, let alone on the solutions. Yet the problems can be explained and solved. The explanation lies in how Shannon's sensible system unraveled, over the course of decades, to become the shambles we have today. It unraveled not because of problems with how individual IRBs function but with how they are instructed and supervised. The solution is straightforward: We must keep committee review but embed it in a reformed system that values both the protection of subjects and the saving of lives.

I DID NOT START OUT AS AN ETHICIST. I am a family doctor, and I spent from 1982 to 1994 in private practice near Seattle. In 1995, I went to Stanford for specialized training in ethics, obtained a JD from the law school, did a fellowship at the Center for Biomedical Ethics, and joined the Stanford IRB.

James Theodore was our IRB chair. The regulations encourage diversity in race and gender; our eleven members included two females and one Asian-American, so we could have done better in meeting this goal. We did have eight doctors from various specialties, a pharmacist, a nurse, and the Reverend P. Kemp Segerhammer, a

Lutheran minister. The regulations require every IRB to have at least one member who is not a scientist and at least one member who is unaffiliated with the institution; the good reverend met both requirements.

Our work was infused with an awareness of history, and we were particularly mindful of the Public Health Service's abusive syphilis research in Tuskegee. We felt that our highest duty was to protect the vulnerable, particularly racial and ethnic minorities. I came to embrace the IRB ethos and to believe that we delayed research or imposed additional costs only when it was necessary to protect subjects.

At each meeting, we reviewed experimental proposals submitted by dozens of scientists. We made the application process straightforward by providing detailed instructions about what we wanted to know about each study's methods, its risks and benefits, and the consent form that would convey this information to potential subjects.

I remember a proposal from Melinda Moir, a young faculty member who wanted to lessen the misery of children in the days after a tonsillectomy. At the time of surgery, she planned to give the child's parent a bottle of pink liquid with instructions to administer one teaspoon every four hours as needed for pain. Half of these bottles contained Tylenol; half contained Tylenol with codeine. She would ask parents to record the child's pain, nausea, and vomiting in the days after surgery. The medication given to each child was assigned at random, and because the bottles were identified only by a number, neither the parents nor the doctors knew which child was getting which medication. Thus, this was a randomized, double-blind study, which made it a reliable way to tease out the pros and cons of each medication in this specific setting.

The regulations governing IRBs spell out the minimum information that consent forms must include, and most IRBs add institution-specific material. Moir's consent form, as we approved it, told parents about the goals of her study, the provision of coded bottles of pain medication, the parent's role in documenting the child's progress, the risks of Tylenol and of codeine, the assignment of medication at random, and the nature of a double-blind study. It also gave Moir's telephone number in case there were any questions.

Her consent form also addressed the remote possibility that the child could suffer harm as a result of study participation. In the 1940s and 1950s, institutions sometimes bought insurance to cover the cost of subject injuries. This practice fell out of favor, and Moir's consent form followed our standard template by warning that if a child were harmed as a result of their participation, "this study does not provide financial assistance for additional medical or other costs." Had we wavered, Margaret Eaton, a lawyer from Stanford risk management and a nonvoting member of the committee, could remind us why this language was not open to debate.

It was not our only precautionary provision. Lest subjects be lured into enrollment by an over-optimistic scientist, Moir's consent form, like every other, stated: WE CANNOT AND DO NOT GUARANTEE OR PROMISE THAT YOU WILL RECEIVE ANY BENEFITS FROM THIS STUDY.

There was more. The consent form promised parents that they would be notified if any new information was learned that might affect their child's condition or their willingness to participate in the study, advised them not to enroll their child in any other research at the same time, reassured them that their child's name would not be mentioned in any published report, let them know that they would not be paid, confirmed that the child would still receive care even if the parent declined study participation, informed them that they were free to withdraw their child at any time, and cautioned them that Moir reserved the right to remove their child from the study. It also gave them the IRB's contact information in case Moir treated them badly, and included the Experimental Subject's Bill of Rights. We had no short consent forms.

That was in the days of paper, and every month I stopped by the committee office to pick up a thick stack of proposals from Stanford scientists. No mortal could have read every submission with care in time for the next meeting, so the IRB gave specific members primary responsibility for each proposal. Once a month, we assembled in a conference room under Theodore's leadership. The primary reviewer summarized the key points of each proposal, and the floor was then open for discussion.

Some IRBs run strictly by Robert's Rules of Order, but that was not Theodore's way. We followed the regulations—we always had a quorum, and Reverend Segerhammer or a substitute who, like Segerhammer, was neither a Stanford employee nor a scientist was always present—but we usually made our decisions by consensus, with voting an afterthought. The regulations authorized us to disapprove research, but I do not remember us ever taking this drastic step. We often approved studies on first review. When something about a proposal troubled us—a confusing consent form, for instance, or a planned enrollment that we thought might be too low to reach statistical significance—we sent it back to the scientist with a request for explanation or modification.

I remember Moir's proposal so well because I told my colleagues, with some heat, that it was unethical to give half the children plain Tylenol. They had just had an operation! Shouldn't they all get a more effective pain reliever, like codeine? At Theodore's suggestion, we tabled the proposal for a month so that Moir could fix this problem. At our next meeting, I learned to my chagrin that most doctors gave only Tylenol, and that if codeine proved more effective it would be a change in practice. We approved the study, and Moir discovered that plain Tylenol relieves tonsillectomy pain just as well as codeine and causes less stomach upset.

SERVICE ON THE STANFORD IRB introduced me to the actual practice of research oversight. We studied scientists' proposals carefully, engaged in thoughtful discussions, and did our best not to obstruct important research. As a law student, and later as a fellow at the ethics center, I also immersed myself in the history and theory of research ethics. I studied with scholars in the field and read their publications. I learned more about the Tuskegee syphilis study, about Southam's cancer cell injections, and about the barbarous research conducted in the concentration camps of the Third Reich. I went to ethics conferences, where people in the hallways told stories about scientists they considered arrogant and IRBs that fought to protect research subjects who were powerless.

In 1999, I moved to Houston and joined the faculty at Baylor College of Medicine, where my new colleagues were scientists. I began going to medical conferences, where people in the hallways told stories about IRBs they considered arrogant that were abusing scientists who were powerless. As I listened, I knew the defenses the IRBs themselves would offer: Scientists cannot judge their own research objectively, and there is no better second opinion than a thoughtful committee of their peers.

But these rationales began to feel flimsy as I gradually discovered how often IRB review hobbles low-risk research. I saw how IRBs inflate the hazards of research in bizarre ways, and how they insist on consent processes that appear designed to help the institution dodge liability or litigation. The committees' admirable goals, in short, have become disconnected from their actual operations. A system that began as a noble defense of the vulnerable is now an ignoble defense of the powerful.

IRB supporters bristle at this criticism. They argue that the helpless victims of Southam and Tuskegee demand eternal vigilance so that such abuses will never happen again. From this point of view, scientists who complain that IRB oversight is too restrictive are placing their frustration on a moral par with the suffering of men with untreated syphilis. But this comparison of scientists who are inconvenienced with research subjects who may die is false. Scientists today are not seeking permission to harm or hoodwink their subjects; they want to use morally sound methods to find new treatments that will benefit us all. Ethics itself is often only a convenient justification for the decisions made by IRBs, since when I press IRB chairs, they often concede that, although they care about ethics, their decisions are ultimately driven by the regulations and the regulatory agencies. As a former federal official has written, IRBs are "more focused on protecting themselves from the regulatory hammer than on protecting human subjects from harm, to the detriment of the clinical research process and, increasingly, to the dismay of the research community."

As a family doctor, I spent thirty years being frustrated by the many times when I could not help my patients because there were simply no treatments available for their conditions. This book grew out of that frustration. It foregrounds my view as a scientist of the problems of restrictive oversight. My goal is to expose the problems that the IRB system creates for medical science and to present a straightforward solution.

MY STORY OF MELINDA MOIR'S TONSILLECTOMY STUDY illustrates, in brief, how an IRB works. I'd now like to look into how IRB review is experienced by scientists, beginning with nephrologist Fredric Coe, who sees patients in the University of Chicago kidney stone clinic.

A patient's first experience of a kidney stone often drives them to the emergency room with severe pain. A urologist may be called in, who will prescribe narcotics and decide whether the stone will pass on its own or if, instead, a procedure like blasting with shock waves or surgery is necessary. In Chicago, once the patient is through with this ideal, they may be referred to Dr. Coe in follow-up.

Coe's primary mission is to help his patients to never have another stone. Coe is also an active investigator; his commitment to research becomes obvious to new patients when they learn that Joan Parks, his scientific collaborator, sits in on every visit, a thick binder of information on her lap. While Coe is talking with a patient, he often pauses to turn to Parks, and they look over published data and trade test results to see whether the treatment plan can be improved.

Coe is one of the more senior members of the Chicago faculty, since he has been treating kidney stone sufferers since 1966. When he started, though, the word "treating" was over-optimistic. Nephrologists at the time had little to offer except waiting and hoping, so although Coe did advise his patients to drink plenty of fluids, his black bag was all but empty. There were pills he could prescribe for headache, high blood pressure, and gout, but he had almost no medications to keep patients who had gotten one kidney stone from getting a second and a third.

Coe is not a passive man, and he began almost at random trying medications that were usually prescribed for other diseases to see whether they might reduce stone recurrence. He noticed, for instance, that some of his patients with stones made of calcium had high levels of uric acid in their blood. Excessive uric acid causes gout and the excruciating joint pain that was perhaps the only thing that Henry VIII, Thomas Jefferson, and Karl Marx had in common. Treatment with allopurinol lowers uric acid levels and relieves gout's symptoms, but Coe had no idea whether it would make calcium stones more or less frequent. Coe tried it, discovered its striking effectiveness, and published his treatment recommendations in the *Lancet* in 1973. They are still standard.

Now in his eighties, Coe is as productive a scientist as ever. I ran a Google Scholar search for "Frederic Coe" as author on May 14, 2020, and it returned 331 published articles and book chapters, often with Joan Parks as coauthor. If you do the same search today, there will be more.

In later chapters, we will see scientists and IRBs locked in struggles over life-saving research. Coe operates in a lower-voltage world, neither saving his subjects from death nor killing them himself. His IRB's interventions into his research are similarly moderate. The board asks for an accounting of anonymous urine samples. It haggles over the wording of consent forms. It requires him to search the literature to produce a bibliography that he must submit as part of his IRB application. It makes him wait for a month, or two, before approving his latest study. None of these impositions is outrageous, but they do add up, particularly since new requirements are regularly piled on the old.

Coe would be less vexed about this oversight if its benefits were not, in his view, so vaporous. The bibliography requirement is a good example. In 1987, Coe began studying the proteins that promote or prevent kidney stone formation, using urine left over from clinic visits. These remnants are anonymous, labeled only with the patient's sex and whether are not they are a stone former. For years, Coe's IRB subjected this research to only the most cursory review. In 2002, however, it began requiring that Coe submit, as part of his application, "a bibliography and summary of recent literature on the subject of urine proteins."

The purpose of this requirement, the committee explains, is to avoid a repetition of a tragedy at Johns Hopkins. In 2001, a research subject, as part of an asthma experiment, inhaled a chemical that turned out to be toxic. She developed a severe cough and, in short order, was hospitalized, put on a ventilator, and died.

A federal investigation concluded that a thorough literature search by the scientist might have found an article reporting that the chemical could be dangerous, leading him to stop the experiment. The investigation also faulted the IRB, for if it had been doing its job, it would have made sure that the scientist had done such a literature search. Because of this tragedy at Hopkins, the precaution of a bibliography was adopted nationally, which is why the Chicago IRB requires literature searches by its scientists, including Coe.

But to what end? Coe agrees that his subjects should be protected from harm, but he is only analyzing their urine. The bibliography is just one item on a checklist for the IRB; there is no reason for anyone on the committee to actually examine it. In a rational world, Coe would appeal to the dean or provost, who would take one look and exempt him from this irrational requirement, but IRBs enjoy what amounts to a presumption of infallibility. This may sound exaggerated, but in fact it is forbidden for a higher institutional official to overturn any IRB requirement for any reason. The dean, who might appoint the committee's chair and every one of its members, is therefore powerless to change its restrictive decisions. This extraordinary arrangement is not just Chicago policy, it is a legal prohibition written into the regulations and applicable to all institutions. And so, to return to Coe, the bibliography, which in the case of his protein analysis is utterly without value, must still be prepared and submitted.

Requirements like this do experimental subjects no good; they only burden scientists and delay their work. Their real harm becomes obvious not in Coe's lab but in emergency rooms down the street and across the country, when the lack of reliably effective interventions makes its mark. When kidney stone prevention fails, patients stagger into emergency rooms begging for morphine. They miss work or school; they require hospitalization and surgery; some who have never used narcotics before become addicts.

I WROTE THAT IRB DECISIONS are "driven by the regulations." But regulations are paper and ink; their implementation, their interpretation, and for all practical purposes their meaning are determined by the powerful federal agencies that oversee IRB operations. This is why Coe's IRB is so anxious to follow federal signals even when they make no sense.

Biomedical research is, among other things, a big business; it is how pharmaceutical companies and biotech startups create new drugs and new devices (like pacemakers). IRBs oversee that research, when it is conducted in humans, under rules established by the Food and Drug Administration.

Flecainide, for instance, was approved by the FDA in 1985 to treat rapid heart rhythms, in which a pulse of 170 or more beats per minute is common. Once it was approved for this specific use, doctors could use flecainide for other conditions without additional FDA review. This is reasonable, for there are dozens of heart rhythm abnormalities and it would be impossible to test each new drug against each arrhythmia for which it might be useful. As a result, cardiologists may try new drugs for serious conditions based on their intuition alone. As you might imagine, this is not always a good thing, but a pharmaceutical company has no reason to invest in research to find out whether its drug is effective in these conditions when it will be used, anyway, even with no evidence whatsoever. The pharma's deep pockets are better used to develop another new drug, research that will likely be done outside of an academic medical center.

This leads to one of the strangest features of medical practice and research today. If a thousand cardiologists use flecainide to prevent rhythm disturbances after a heart attack, they are under no obligation to keep track of the results. They are just practicing medicine based on the usual combination of evidence and intuition. If, on the other hand, a dozen cardiologists want to treat some of their patients with flecainide, give a placebo to the remainder, and keep track of the results, then they are proposing research and must first get permission from an IRB. This IRB will operate not under the rules of the FDA, for whom this drug has no further interest, but under the rules of the Office for Human Research Protections, an office within the federal Department of Health and Human Services. Similarly, OHRP rules governed the Stanford IRB as we reviewed Moir's comparison of Tylenol and codeine.

Research that has nothing to do with medical drugs and devices—research like Coe's investigation of urine proteins—is also reviewed by IRBs operating under OHRP's rules. This is always true when the research is federally funded, and sometimes (this is in flux at the time of writing) for other funding as well.

The interactions between the FDA and Big Pharma are complex, high-dollar affairs that could be much improved, but that is not our concern in this book. We will occasionally consider research overseen by the FDA, but our focus will be on science that is under the supervision of OHRP, usually conducted in academic settings. When I refer to the IRB system, I have in mind the network of local IRBs plus their federal supervisors. Federal oversight is intermittent at best, since OHRP does not have the resources to closely supervise thousands of IRBs. But an IRB that catches OHRP's eye may be in for a very hard time.

Director Shannon operated NIH's own CRC in a reasonable and efficient manner. He no doubt assumed that, after the IRB system was created, federal supervisors and local IRBs would cooperate in pursuit of their shared goals. In 1971, an official guidebook spelled out the government's expectations of how IRBs should discharge their duties. It suggested a collaborative approach, noting that "the decisions required by this policy must depend upon the common sense" of IRB members.

It is grounds for concern, if not alarm, when officials feel they need to endorse a quality that is one of the (generally unspoken) foundations of rational government. In any case, federal support for common sense did not survive. Its demise was triggered by a question in a Capitol hearing room in 1998, where members of Congress were grilling Gary Ellis, head of the Office for the Protection from Research Risks, the predecessor of OHRP. The agency's duties included authorizing IRBs to review federally-funded research. It had the power to suspend that authorization and require all research that had been approved by that IRB to stop. Such a suspension was Ellis's hammer, since it would shut off the flow of federal dollars—often millions of dollars per day. How long was it, the legislators asked, since Ellis had shut an IRB down? He admitted that it had been years. In response, the legislators expressed deep misgivings about his fitness for a government salary and a corner office.

Ellis went back to his office and took the hammer off the shelf. The easiest targets were institutions whose IRBs made what Ellis considered a major error. Johns Hopkins had responded to the asthma study fatality by immediately shutting down all of that scientist's research and launching an investigation. Ellis went further, temporarily cutting off all federal funding for thousands of experiments that had nothing to do with asthma. Work could begin again, and funding could be turned back on, only after the Hopkins IRB re-reviewed every protocol it had previously approved.

OPRR and, later, OHRP imposed similar punishments on a dozen other institutions. One other case involved a death. In many of the rest, nobody had been injured or even placed in harm's way; the issue was one of policies that the agency deemed inadequate, or even of a failure to document that policies had been followed, and the standard apparently was perfection, with every box checked. In this time of high anxiety, medical schools, hospitals, and universities added IRB staff and tightened committee procedures to avoid being the next victim. IRBs that had once focused on avoiding injuries to subjects now anxiously sought to avoid the federal hammer.

The legislators who prompted this reaction may have been pleased. Other people, both in and out of government, felt that these draconian penalties had gone too far, and Ellis was soon forced out. The spate of punishments eased in 2001, but OHRP continues to periodically remind institutions of its power by enforcement actions that are often, in the eyes of some observers, unjustified. IRBs take the federal threat very seriously, and they continue to do their best to check every box, whether it makes sense or not. They would rather be foolish than appear careless.

IRBs can demand that any aspect of a proposed experiment be changed. Their exercise of this power is supported by ethicists, who see a moral dimension in every detail of experimental method. Consider, for instance, the number of participants a study plans to recruit. This might seem like an unimportant detail, but no aspect of experimental method matters more to the scientist. An investigator who tries to sign on too many participants will run out of time and money. Enroll too few, and the results will not be statistically significant, leaving the scientist with nothing to show for all of that time and effort. So before the first subject is ever approached, the scientist sits down and reviews the funds on hand, the time available, the number of people eligible for the study, the expected benefit of the new treatment, and a plausible margin for error, then swallows hard and decides how many people to enroll.

Most IRBs include active researchers. Some are careful to respect investigators' plans unless there is an obvious problem; others are liberal with suggestions for improving the studies they review, which may include changing the enrollment target. They may believe, for any one of a dozen reasons, that it should be higher or lower, and the committee may vote to return the proposal to the scientist with a recommendation for that change.

And what does a "recommendation" mean? There is no single answer. One IRB takes no offense if the researcher rejects its recommendation about study design; when another says "we recommend," they really mean "we insist." In either case, sometimes the committee is right and the scientist is grateful for the advice, and sometimes the committee is wrong.

The ethicists' argument is that enrolling too few subjects is a moral wrong, since their time and exposure to risk is wasted if the result does not reach statistical significance. Enrolling too many is also a moral wrong, since more subjects are exposed to risk than was necessary to obtain a result. I think this reasoning is correct. Where I part company with the more activist IRBs is in how the correct number of subjects should be determined. Activist IRBs sometimes believe that their lesser knowledge of the specifics of the area makes them more objective, and their consciousness of the moral dimensions of enrollment makes them more ethical, and so their view of the matter should prevail. I disagree. Unless the case is exceptional, I believe that the scientist, who is an expert in the area, has developed the plan, and has obtained the funds, should have the last word.

Another factor adds to the problem of IRB disrespect. Training materials encourage IRBs to see researchers as people with a conflict of interest, since a successful experiment may lead to academic promotion, profitable stock options, and more. This does happen, although less often than IRBs imagine or scientists wish. But this training encourages IRB members to see scientists as corruptible and to consider themselves to be morality's only defenders. This general distrust of researchers emboldens some IRBs to require modification of enrollment targets. But limitations of time and money independently drive the scientist to the same enrollment goal as the ethicists—just enough subjects and no more.

Fredric Coe eventually became so outraged by IRB oversight that he wrote an article describing his experiences; many other researchers have done the same. Not all scientists oppose the system; some consider its requirements a small price to pay for the important goal of protecting subjects. But even those who oppose it rarely take the time to speak out. They have experiments to conduct and, they hope, lives to save.

They live with IRB review as best they can. The team with enrollment troubles agrees to raise (or lower) its target and submit a revised proposal. It is too late for the February meeting; the committee will consider it in March, although that meeting will be delayed for spring break. Another team wants to survey doctors to find out whether they are ordering mammograms appropriately. The IRB-required consent form will ominously inform the doctors, "This medical center will not provide treatment for any injuries you incur as a result of completing this questionnaire." This warning will make the survey appear comical, but there is no negotiating with the IRB. If the scientists go to a conference hoping for sympathy, they may meet investigators from another university whose IRB worries that answering a questionnaire on post-traumatic stress disorder might move people to suicide.

Scientists are frustrated when they learn that they cannot appeal IRB decisions to institutional officials. They are bitter as they remember experiments that were so restricted by IRB requirements that they were never completed. They become cynical about an environment in which it is unethical to enroll too few subjects, unethical to enroll too many, and only the IRB can tell the difference. They feel defeated when they realize that their IRB is unreasonable not because it has lost its common sense but because it fears federal punishment, which means that real improvement will require an act of Congress.

IRBs have great power. They are under tremendous pressure. They are not widely loved. In this difficult situation, some experience a regrettable loss of empathy for those they regulate. I remember one researcher who had laboriously assembled the funding and manpower needed to do a useful and risk-free study. The IRB demanded changes that would raise his expenses to an impossible level. When he phoned an IRB official and explained his dilemma, he was told, "Cost is not our concern."

His experience is not uncommon. In 2003, I conducted a small email survey of NIH-funded scientists. One, whose access to stored materials had been restricted by IRB decisions, wrote, "I hope all those at OHRP and the ethicists die of diseases that we could have made significant progress on if we had these valuable research [materials]."

OHRP sometimes moves beyond reviewing individual IRBs to directly supervising scientists. In 2002, the agency attacked the methodology of a major study of how best to treat patients on ventilators. Two outside panels eventually found OHRP's objections to be baseless and the study resumed, but only after a delay of almost a year.

OHRP is well aware of its unpopularity with scientists, and in 2017 and 2018 it announced changes in the regulations. These resolved some longstanding problems, but much of the system remains as it was. Coe has experienced no improvement, writing in 2020 that he had to obtain full IRB review, lasting months, in order to collect unidentifiable urine samples from normal people. In his essay, Coe doesn't call IRB oversight oppressive or rigid. He calls it a tyranny, which is how oversight feels when it doesn't work. I think it would be more fair to call it broken. I have no wish to demonize conscientious IRB members who are doing their best. Nor do we need to, since the system's failure has a cause that can be identified and a solution that, when implemented, will still rely on local committees to do its work.

WE TURN NOW TO THE QUESTION of just how the system became so dysfunctional. The simple answer is that, because it does not recognize the moral value of better medical treatments, the IRB system tries too hard to eliminate subject risk. You might wonder how that could possibly be a mistake; we all know that eliminating risk is impossible, but shouldn't all institutions, including IRBs, try? It turns out that while reducing risk is a good idea, trying to completely eliminate it is not. The reason is not intuitively obvious. It reflects the law of unintended consequences, and it was discovered only through bitter experience.

In the 1960s, institutions of every kind grew weary of devastating losses from risks that seemed obvious in hindsight. In response, they painstakingly identified their greatest risks, established policies to eliminate them, and assigned people to enforce those policies. As it sought to banish risk to experimental subjects, the IRB system was part of this movement. For all of these institutions, eliminating risk seemed simple prudence, yet corporations, churches, universities, governments, and nonprofits all found that this logical approach could lead to a fresh hell of disasters.

Analysts picked through the wreckage and discovered a striking pattern: Attempts to extinguish one risk invite disaster from a second, hidden, hazard. When the Forest Service sought to eliminate forest fires, flammable brush accumulated and eventually led to unprecedented conflagrations. Suppressing the smaller, earlier fires had seemed the only reasonable policy, but the real choice was between a smaller fire then and a larger one later. Across a variety of settings, attempts to eliminate risk led to equally dire results. Rather than attempting absolute suppression, risk is better managed by balancing the obvious hazard with its inconspicuous but equally dangerous twin.

Risk will never be fully eliminated, and modern methods of risk management can still fail spectacularly, but they are the best we have, and they have been embraced by institutions of every kind. Applying these lessons to research oversight, the IRB system should recognize that it needs to consider both the protection of research subjects and the public benefits of research. This balancing approach is standard in every other area of risk management, but the IRB system operates in a backwater, its methods half a century behind contemporary norms. Only by balancing subjects' need for protection with society's need for better treatments can it help scientists develop the treatments we all need. This the current system—IRBs and their federal supervisors—does not do.

It is not the fault of the IRBs themselves, who are entirely capable of this balancing. They fail to do it only because ethicists, OHRP, and the relevant regulations all urge them to disregard the public's need for research. When these counterproductive forces are replaced by a sensible system of risk management, conscientious IRBs, like the one I served on at Stanford, will be able to make sound decisions.

Because misguided oversight is baked into the system, reform will require fundamental change at every level. Congressional action to reform the system is urgently needed and long overdue. In the absence of Congressional action, ethicists and scientists are debating smaller changes that might help with some of the system's worst habits. That debate is taking place out of view of the public, for it is conducted largely in books that are aimed at scholars and in journals that are behind a paywall (a year's subscription to the *American Journal of Bioethics* costs $270).

I want to take you behind that paywall. We will review the system's contested history, its troubled present, and the better future that requires only an act of Congress. That's not too much to ask for a system with great power over the medical research that should be saving more lives. In this book, we will focus on specific people and events that illustrate the more general trends. We will begin with a young woman whose modest sacrifice in the interests of medical research has never been recognized.

# 2

# The Birth of Research Oversight

PATIENT R.M., NINETEEN YEARS OLD, was lying on a gurney in Massachusetts General Hospital, waiting her turn for routine surgery. Two attendants came to her side, explained that they didn't want her to fall off the gurney after she became unconscious, and fastened straps loosely across her chest, hips, and legs. It was all routine, and although our information is patchy—the year was 1948 and the records are skimpy—she apparently sailed through surgery. She would never know that while she was asleep she helped scientists understand the effects of ether on kidney function.

Ether anesthesia was first demonstrated at Mass General in 1846, when William T.G. Morton used it to allow the painless surgical removal of a young man's neck tumor. In 1949, about a century later, the hospital was a center of anesthesia research, often led by Henry K. Beecher, who was chief of service. Beecher, a confident man with thick hair and a handsome, square face, had wide-ranging interests that included the form of kidney failure that can follow surgery.

In a typical case, an auto accident victim is brought to the operating room with mental confusion, low blood pressure, and low urine output—all signs of shock due to internal hemorrhage. The surgeons give him blood transfusions and find and fix the source of the bleeding. After the operation, he wakes up; his blood pressure is normal, and he should be on his way to recovery, but there is not a drop of urine. Kidneys are particularly susceptible to damage from shock, and his have been irreparably damaged by the low blood pressure, the lack of oxygen, and, perhaps, the anesthetic (it can be difficult to untangle the side effect of an anesthetic from a patient's other problems). Today a patient with postoperative kidney failure would be started on dialysis and sent home. In 1949, if his kidneys didn't recover on their own, he would die.

Before World War II, deaths like this aroused little experimental interest. Few medical school faculty members, with heavy practice and teaching loads, had the time and resources to seek ways to reduce the toll. But a postwar surge in government funding, much of it from the National Institutes of Health, made it possible for Beecher to pull together teams to investigate important clinical problems like this one. Was it possible that ether anesthesia was making postoperative kidney failure more likely? There was a newer anesthetic available, cyclopropane—would it be easier on the kidneys? Beecher's team planned an experiment in which a series of patients would be given either ether or cyclopropane and their effects compared.

Beecher did not ask R.M. whether she wanted to participate in this experiment. He simply went ahead, which is what most scientists were doing in 1949. Looking back on that era, hematologist David Nathan remembered that he applied a rough version of the Golden Rule to experiments: He would never do anything to a research subject that he would not do to himself. Once an experiment passed this threshold, however, his idea of informed consent was to say, "You are the patient. I am Doctor Nathan. Lie down."

Even this whisper of consent was unnecessary with R.M., who was already lying down and would soon be unconscious as well. She was wheeled to an induction room, and Beecher, or one of his colleagues, placed an ether mask over her nose and mouth. Several attendants were standing near; the moment her eyes fluttered shut, they cinched the straps, binding her tightly to the gurney. Her breathing deepened and then turned to heavy gasps while her entire body struggled vigorously (one of ether's predictable and fortunately temporary effects); the attendants worked hard to make sure the straps held and she did not injure herself. Her breathing slowed, her body relaxed, and ordinarily the surgery would have begun.

But Beecher's team had other plans. As R.M. was kept under the ether, they administered two chemicals that help measure kidney function and then collected blood and urine samples at carefully timed intervals. This was not a rapid process, for they wanted to be sure their results were accurate. Forty-five minutes passed before they had all the data they needed and the surgeons could begin the operation. R.M. had made her contribution to science, and taken the risk of extended time under anesthesia, all unaware.

This process was repeated with fourteen other patients. After analyzing the data, Beecher's team found that ether and cyclopropane both reduce blood flow to the kidney, but the drop with cyclopropane is sharper. In their published report, they suggest that, especially for patients who are in shock, ether is the safer anesthetic, a recommendation with the potential to save many lives.

Beecher eventually came to feel that he had been wrong to subject R.M. to experimental risk without her consent, and that other scientists were wrong to use similar practices. Within a decade, Beecher began a campaign against these abuses, seeking to persuade the medical establishment to act with more concern for experimental subjects' rights and welfare. The stakes were high: If he failed, officials wielding federal power might step in.

JAMES SHANNON, WHO HAD BEEN AT THE NIH since 1949 and was to become one of its most celebrated directors, would play a crucial role in the next stage in the story. He had trained at New York University as a kidney physiologist. With the outbreak of World War

II, the government pressed him into work as a scientific administrator.

One of Shannon's biggest challenges was a critical shortage of medications for malaria. After the Japanese bombed Pearl Harbor, it became apparent that American troops would soon be in the jungle, fighting an enemy that would keep quinine, the only effective antimalarial, out of Allied hands. Shannon coordinated a nationwide effort involving scientists at many institutions to find and test new drugs against the disease. Some of this testing exposed subjects to risks that would be viewed as unethical today, but a war was on and the lives of millions of soldiers and sailors were at stake. And it worked: Shannon's researchers found several promising medications.

In addition to its scientific achievements, the wartime work demonstrated that generous federal funding, aimed at specific medical threats, could save lives. After the war ended, Congress began to ramp up funding for research into peacetime health problems like heart attacks, while the military sponsored investigations into diseases like hepatitis that were a problem for soldiers and civilians alike.

Shannon was in charge when the NIH opened its own hospital, the Clinical Center, in 1953. A photograph from these early years shows him smiling, with tortoise-shell glasses and a careful pompadour, one foot on his desk. The casual pose belied the serious challenges he faced, including the risks posed by his own scientists to their experimental subjects. The diseases under investigation were life-threatening, but so were some of the treatments being developed. The surge in federal funding made possible all manner of experimentation, the scientists limited only by their own conscience. As one of Beecher's biographers wrote, breakthroughs like the polio vaccine "had combined with a flood of federal research money to generate excitement and a sense of urgency, a feeling that anything was possible and that delays would be measured in human lives."

Shannon saw that an oversight system was needed, and he established the Clinical Research Committee. The CRC comprised leaders from the NIH's institutes, and it reviewed all proposed research involving either unusual risk or healthy volunteers. Shannon's decision to vest a committee with the responsibility for research oversight is worth highlighting. He could have put a suitable individual in charge, perhaps the scientist's department chief, but he preferred a committee, which provided a broader range of experience and perspectives.

CRC review began by evaluating the proposed experiment's level of risk. Research that was especially dangerous—inducing an infection that might be fatal, for instance—was rejected, a judgment that ended at least two proposed experiments before they got off the ground. Once the committee found that the risks were acceptable, it decided what consent process to require, for subjects had a right to know what they were getting into. In the end, most proposed research protocols were approved, for the CRC's goal was not to eliminate risk, which would require closing the Center, but to balance the potentially competing interests of subject and society. The committee reported to Shannon, who could overturn its decision if a scientist made a persuasive appeal, but this seems to have happened seldom or never.

The CRC review system worked. During this era at the Clinical Center, no subjects died, no scandals rocked the agency, and the work of finding new cures for heart attacks, lethal infections, and much more moved smartly ahead. This foray into the social control of scientists was one of the NIH's most successful experiments.

But the CRC's control ended at the NIH's gates. In institutions across the country, thousands of scientists were spending millions of dollars of federal money with few if any strings attached. By the late 1950s, a few medical schools had committees to look after subject welfare, Harvard among them (Beecher served as its chair). But most institutions showed little or no concern about the threat of abusive research. Senior NIH officials regularly advised investigators and institutions about how to reduce the risk to subjects, since Shannon did not want to have to explain to Congress why his beneficent medical machine had injured or killed an experimental subject. But, for the time being, federal officials felt no imperative to impose controls to reduce risks beyond the NIH campus.

LIKE THE NIH, THE US MILITARY invested in medical research, particularly into hepatitis and other communicable diseases. Hepatitis is mild in children but can be severe in adults; during 1943–1945, it was the most serious cause of disabling illness and one of the most common fatal diseases among American troops. An official Army history recalls that military doctors were confronted with "an enormous number of cases of a disease for which the means of spread were not clearly understood, no methods of prevention were known, and no specific therapy was available."

Beecher served in the Army during the war and saw the impact of hepatitis first-hand. After he returned to his operating room in Boston, whenever he administered blood transfusions, he knew that some contained undetectable contamination that would cause fulminant hepatitis. Nationwide, thousands died from infected transfusions every year; hepatitis due to other causes killed thousands more. Both military and civilian authorities were anxious to better understand the disease.

A Chicago orphanage with a serious hepatitis problem presented an opportunity for investigation. Pregnant women who were unable or unwilling to care for a child would move in until they delivered. Their babies would be separated from their mothers at birth and cared for by nurses and nurse trainees during stays that might last up to three years. Although nurses on the permanent staff were largely spared from hepatitis, student nurses developed the disease at an alarming rate. Soon after they arrived, these trainees got sick and went home, unable to work, for weeks or months. Adult hepatitis is a truly miserable disease; even people with mild infections often complain they'd rather be dead.

Every Chicago student nurse's training included outside rotations, like that at the orphanage. Those who worked at other institutions generally completed their obligations without incident, but, one after another, student nurses at the orphanage came down with hepatitis. This was hard on the orphanage's staffing, and on its reputation as well. The director appealed for help to pediatrician Joseph Stokes, whose work on hepatitis was being funded by the Army.

To keep the trainees from falling ill, Stokes first had to find the infection's source. The students were healthy when they arrived and became sick soon thereafter; could the food they were eating be infectious? No, since everyone at the orphanage ate the same food, and few of the other residents and staff got sick. Stokes concluded that the students must be catching the infection from one of the other groups in the orphanage: pregnant women, orphans, or the long-term medical and nursing staff. It was something of a locked-room mystery, for the babies and the other adults all seemed quite healthy.

After considerable fumbling, much wasted time, and more than one dead end—the process we call science—Stokes found an important clue. Some of the infants, some of the time, had abnormal liver function tests. Stokes checked these tests only because hepatitis is a liver disease and he reasoned that somehow, somewhere, he should be able to find its footprints. The infants showed no other sign of illness—they were healthy and gaining weight normally. But if the liver function tests were to be believed, they also had an invisible infection, in a form that was too mild to bother the infants themselves, but virulent enough to land the student nurses flat on their backs in bed.

Stokes probed further. There were a lot of babies in the orphanage, and the student nurses were changing a lot of dirty diapers. They were supposed to wash their hands carefully after each diaper change, but Stokes noticed that they were not always scrupulous in following this requirement. Perhaps, he speculated, the stool was infectious. If so, a particle of stool that had not been washed off a nurse's finger might make it onto the sandwich she ate for lunch. An interesting speculation, but not the kind of proof that would move science forward.

For proof, Stokes would need to show unequivocally that the babies' stool could carry the disease. The nurses would not be eager to participate in such an experiment, but Stokes had a contact in a local correctional facility. This experiment would substitute a prisoner downing a contaminated drink for a nurse eating a contaminated sandwich.

Stokes made an edible preparation including stool from the infants with abnormal liver function tests. His published report does not contain the recipe, but if I were in his shoes, I would use a small drop of infected stool in a large chocolate milkshake. Stokes persuaded ten prisoners, whom he describes as volunteers, to partake of his concoction. Out of those ten, two developed hepatitis (the other eight may have been immune from previous infection). This result confirmed that the nurse-trainees were contracting the disease by changing the diapers of infants who had no symptoms but were nonetheless infected. Stokes told them to wash their hands more carefully. The students did as they were told, and just like that, the problem disappeared.

Hepatitis was a problem anywhere hygiene was poor, including the Willowbrook State School on Staten Island in New York. Willowbrook, which housed children with intellectual disabilities, was underfunded, overcrowded, and brimming with infectious diseases of all kinds, including measles, which frequently led to lethal outbreaks. Hepatitis was endemic—90 percent of the residents eventually contracted it—but it produced only mild symptoms, or none at all, in the children. Here, as in Chicago, its burden fell mainly on the nurses.

In the orphanage, the babies' infectious stool was largely contained by diapers. Willowbrook's children sometimes wore diapers but it made little difference, since they were old enough to play with their feces and soil their walls, floors, and playmates. Even meticulous hygiene could not protect the nurses in this heavily contaminated environment, and they regularly fell ill. In 1956, Willowbrook's director asked for help from pediatrician Saul Krugman, a junior faculty member at New York University.

In order to focus on hepatitis, Krugman set up a separate unit, one that was less crowded, better staffed, and largely free of the infections that were prevalent elsewhere in the institution. The only common infection in this ward was hepatitis, and the children caught it because Krugman gave it to them. He gave it to them either by injection or orally, probably using some variation of the milkshake method. This may seem shocking, but bear in mind that many of the children were incontinent and sucked their soiled fingers. The ingestion of stool was therefore an everyday occurrence. Krugman provided extensive information to the parents and enrolled children only with their permission.

In a 1959 report, Krugman reviewed the induced hepatitis of R.J., a six-year-old boy who was given "Willowbrook virus" with parental consent. That was on day zero, and over the following months, under the careful watch of the hepatitis unit staff, R.J. showed no sign of disease. Yet he had not only acquired hepatitis, as indicated by abnormal liver function tests between days 26 and 66, he was infectious. This proved that children at Willowbrook with hepatitis could infect others even if they felt and looked fine. Extending the result to the population at large, Krugman explained that someone with early disease and no symptoms could infect others "by intimate contact" or "by donating blood or by contaminating needles, syringes, dental instruments, or other articles." He concluded that "the person with subclinical disease is probably the most active and efficient disseminator of the virus." At Willowbrook, healthy-appearing children like R.J. were spreading the disease to other children, who usually had mild disease, and to the nurses, who became seriously ill.

KRUGMAN'S REPORT WOULD CATCH BEECHER'S EYE. When he was not conducting experiments, like his study of ether and kidney function, Beecher worked in the Phillips Pavilion, a luxury annex to Mass General, where he provided anesthesia to Boston's elite. He was busiest at the beginning of surgery, shepherding the patient into unconsciousness, and at the end, reversing the process, but he had much less to do while the surgeons were operating. A committed multitasker, he made the most of this time of lower stress, tending to the patient when needed but generally focusing on his academic obligations. If the patient was not breathing spontaneously, Beecher would squeeze the ventilating bag with one hand and read journals or correct proofs with the other.

Around 10:30 a.m., his secretary, Ruth Studley, would arrive to give Beecher his morning mail and take her position on a stool just outside the open operating room door, where she would take his dictation. He did not worry that his patients might return to consciousness during the surgery; a colleague remembered that he liked his patients "very asleep, with pupils dilated, if not wide enough to qualify as open windows to eternity."

In this way, while continuing his busy personal practice, Beecher kept up with medical journals detailing advances in every area of medicine. There were important discoveries in hepatitis, cancer, and heart disease. Penicillin, the wartime miracle, had been joined by other antibiotics, and scientists worked to discover the best combination of drug and dose for each infection. Experiments like these transformed clinical medicine and enabled doctors to save countless lives.

But as he leafed through these reports, Beecher was troubled by the willingness of some scientists to gloss over the risks taken and the sacrifices made by the experimental subjects, often called "volunteers." Sometimes these volunteers were said to have been informed of the experiment's methods and goals, and to have given their consent. Beecher suspected that this was often a lie. Scientist James Scott, for instance, had proved that a high dose of the antibiotic chloramphenicol suppresses the bone marrow. Beecher was sure that no reasonable person would knowingly volunteer for such an experiment. Another article described the serious complications that followed when penicillin was withheld from servicemen with streptococcal infections. In a third, charity patients with typhoid fever were found to be more likely to die if they were given a placebo instead of chloramphenicol, a known effective treatment. Experiments like these troubled Beecher deeply.

The editors who published these articles, and the doctors who read them, seemed to give no thought to the silent sacrifices of the people behind the data. Beecher did, and he was appalled, but he was powerless to effect change. He held a professorship at Harvard, but his authority did not extend beyond the section of anesthesia, which was a small part of the department of surgery. Formal power over biomedical research rested in the hands of medical school deans, journal editors, and the heads of medical societies. These people had built their reputations on experiments of just this type, and they saw nothing wrong in conducting research that jeopardized the welfare of a handful of subjects if the results might save many lives in the future.

Beecher considered this reasoning to be doubly wrong. He emphasized the immorality of making vulnerable patients martyrs in the cause of science, but that was not all. He was also keenly aware that abusive research could pose a threat to the profession. Beecher, like Shannon, saw what most of their colleagues were blind to: The public that had accepted the need for hazardous research during the dark days of the war now felt that the time for sacrifice was over. In the post-war climate, an abusive experiment might cause a scandal that would lead to government regulation—something Beecher sought to avoid at all costs.

He began pulling together materials for a major article, one that would bring the problem of abusive research to the attention of the profession and to present his preferred solution.

Beecher began by compiling relevant policies from the United States and abroad, writing medical organizations and government agencies to ask whether they had rules for research. He received answers from France, England, and the United States. He considered them all unsatisfactory.

The best-known guideline was the Nuremberg Code. During the war, Josef Mengele and other Nazi doctors had experimented on concentration camp prisoners without concern for their suffering and death. Mengele escaped to Argentina after the war, but other doctors were tried at Nuremberg. The tribunal's judges, with an assist from Andrew Ivy of the American Medical Association, drew up a list of ten ethical obligations of the medical researcher, asserted that these were universally valid, found the Nazi scientists to have violated them, and sentenced seven of them to hang. These obligations became known as the Nuremberg Code. They were a milestone in research ethics, and they are still cited with respect.

Beecher described the Nuremberg Code as "a valiant attempt to codify permissible experimentation in man." The first point, which stated that "The voluntary consent of the human subject is absolutely essential," struck Beecher as "simple, straightforward, and absolutely to the point." It identified one aspect of the barbarity that was Nazi research. But Beecher, on reflection, saw many difficulties in applying this principle to contemporary experimentation. What about research into better treatments for major mental illnesses, whose victims were incapable of consent? The Code made no provision for consent by a guardian.

Another point of the Code required the investigator to inform the subject of "all inconveniences and hazards reasonably to be expected." Beecher felt that this reasonable-sounding requirement could not be met, for early in research even major problems may be unknown. He saw every other point as similarly flawed. The code, in his view, was a well-meaning attempt to create simple rules for a complex experimental universe, one that was doomed to fail.

Beecher then reviewed the opinions and policies of various authorities, including Pope Pius XII, the American Medical Association, the French academies, the World Medical Association, the British Medical Research Council, and the United States Public Health Service. Beecher did not know of the PHS's experiment in Tuskegee, which had then been underway for almost a quarter of a century. Even so, he doubted that the PHS or any other organization had an answer for the ethical conundrums posed by research. He concluded that codes were appealing but useless, and likely in most cases to do more harm than good.

Beecher did see a path to reform, one that did not rely on a code or a set of guidelines. First, he felt that research risk should be stratified as absent, present, or present and excessive. If there is no risk, then there is no need to provide potential subjects with any information. If risk is present but not excessive, prospective subjects are entitled to a full explanation, so that their consent, if they agree to proceed, is valid.

Beecher was angry about scientists who enrolled supposedly consenting volunteers in excessively risky experiments. He pointed out that patients will agree to almost anything their doctor suggests, so it would be wrong for a scientist to assert that a dangerous experiment was proper just because there is a signature on a form. Excessively hazardous experiments should never be conducted.

But who decides whether risk is "excessive"? This is a critical point, since a scientist and an outside observer might disagree as to the amount of risk in a particular case. Beecher noted that scientists must recognize that they may not be objective about the risks of research, and he implored them, when in doubt, to consult with a suitable committee. He pointed out that this process was similar to the system that Shannon had put in place at the NIH's own Clinical Center. But in contrast to the NIH's system, Beecher's consultation process would be entirely voluntary—he wanted to improve scientists' ethics, not to infringe their autonomy.

Beecher's essay was forceful, but it was also diplomatic, since he wrote not a word about the many specific experiments that fueled his outrage. To the contrary, he stated flatly, with what sincerity we cannot know, that the unethical experimentation he was aware of was not due to "willful or unscrupulous" behavior but rather was the result of "ignorance or thoughtlessness."

The medical establishment happily endorsed this call for higher ethical standards, which they regarded as having many virtues. Beecher did not identify specific unethical experiments; he savaged codes and guidelines, which threatened to make research more difficult; and he opposed mandatory oversight. There was much to like, and so this article was "authorized" by the AMA's Council on Drugs and "adopted" by the AMA's Committee on Research.

The medical leaders who were happy to thank Beecher apparently thought that abusive experimentation was being conducted somewhere else by someone else, since, in practice, his exhortations left experimental practice untouched. Beecher's vigorous attack on formal codes amounted to a plea that American medicine manage itself better to prevent the calamity of outside regulation. But few if any of his colleagues shared Beecher's sense of urgency.

At the NIH, Shannon's approach was more pragmatic. He recognized abusive research as a serious problem, and he took a methodical approach, first seeking to understand the scope of the difficulty, then pondering what regulatory tools might be appropriate. In 1960, he asked the Boston University Law-Medicine Institute to learn what steps, if any, hospitals and medical schools were already taking to protect subjects. He found that few institutions felt that ethically defective research was an important problem, and that even fewer were doing anything about it.

THE READINESS OF SCIENTISTS to perform experiments with little concern for their subjects' safety was part of the reason that science in the 1950s and 1960s was progressing so rapidly, with innovations in antibiotics, vaccinations, surgeries, and much more. That period also included experimentation that studied the fundamental secrets of cancer, research whose seeds had been sown in Chicago almost half a century earlier.

Surgeons rarely keep part of a cancer for their own use, but Nicholas Senn (1844–1908) was never bound by custom. While other scientists followed their plodding ways, Senn would move directly from his best guess to the daring experiment that would prove it. In addition to being an innovative researcher, he had been chief surgeon of the Sixth Army Corps during the Spanish-American War and president of the American Medical Association from 1897 to 1898.

Around that time, some of Senn's colleagues proposed that cancer was an infection. The idea was an extrapolation from the discoveries of Robert Koch. In the 1870s, Koch had taken blood from a sheep that had died of anthrax, a feared and little-understood disease, and injected it into a mouse. The mouse died the next day. There was something fatal in the sheep's blood, something that could be transferred. Through careful laboratory work, including more animal transfers, Koch eventually proved that anthrax was caused by a rod-shaped bacterium. He later identified the causes of two other major infectious diseases, cholera and tuberculosis. If cancer, that fearful killer, could also be shown to be a form of infection, a new world of possibilities would open.

Scientists did not dare try to transfer cancer from one human to another. When, in laboratories around the world, they attempted to transfer human cancer to goats, dogs, cats, rabbits, guinea pigs, rats, and mice, they repeatedly failed. But this did not prove that cancer was not an infection—it was still possible that a human-to-human transfer would confirm infectivity.

Senn was sure that cancer was not infectious, and he meant to prove it. The best place to do it was in his second home—the operating room—and it was there, in 1901, that one of the strangest scenes in surgical history unfolded. Senn had recently been consulted by an Irishman with an aggressive cancer of the lip; the lymph nodes under the man's jaw bulged with metastatic disease. In the operating room, Senn removed the diseased tissue, but instead of sending it all to the pathologist he took one of the lymph nodes—firm, moist, and malignant—and cut off a piece the size of a split pea. Then someone, either an assistant or Senn himself, made a small incision in Senn's upper forearm, carefully inserted the glistening sliver of tissue, and sewed the incision shut. Cancer might not be transplantable to goats or guinea pigs, but if it were an infection, it would certainly flourish when implanted in the perfect host. A human. Himself.

Over the next day or two, Senn's forearm developed a small swelling; within a week, a firm lump the size of a pea had formed. A lesser man would have asked a colleague to cut out this menacing growth. Senn, however, remained tranquil, and the swelling soon disappeared. Today, such an experiment would receive widespread media attention. But that was a time when news of medical research, with its triumphs, failures, and follies, was largely confined to scientific journals, and it was in such a journal that Senn announced his discovery.

We know now that while Senn believed that he had proven that cancer is not an infection, he had done no such thing. He had actually provided an example of how the immune system uses a flood of specialized cells and proteins to attack foreign tissue, regardless of whether it is a bacterial infection, a stranger's cancer, or something else. The vigorous battle being waged between his body and the cancer caused the swelling, which melted away once the war was won.

Fifty years later, Chester Southam tackled the same problem with a similar, although far more sophisticated, approach. Senn had shown that the body could reject cancer, but his experiment shed no light on the details. Southam, then a junior scientist at Sloan Kettering Cancer Institute, wanted to learn more. How does this rejection take place? What are the steps? What would augment or diminish the process? Are some people better able to reject a transplanted cancer than others?

Southam began his experiments in the 1950s. Just a decade before, an experiment involving the systematic manipulation and transplantation of cancer would have been impossible, for scientists could not obtain fresh tumors from surgeons on demand. Other long-standing questions about health and disease had gone unanswered because it was so difficult to keep cells, cancerous or otherwise, alive for experimentation.

George Gay solved this previously intractable problem using a portion of Henrietta Lacks's biopsy. Her cancer cells, now named HeLa, opened new avenues of research, including Southam's investigations. By this time, Senn's demonstration of the body's immunity to foreign material had been replicated in many contexts, and scientists knew that people usually reject transplants from anyone except an identical twin. A patient with extensive burns, for instance, benefits from skin grafts from an unrelated donor not because the grafts take—they don't, because they are attacked by the immune system and soon slough off. But they temporarily protect the body as it makes its own lasting repairs.

While Senn had not hesitated to use himself as a subject, Southam was more cautious. Although he knew that injected HeLa cells would probably be attacked and destroyed by the body's immune defenses, he also knew these were aggressive cells that had killed Lacks herself. Southam therefore chose, as his first experimental subjects, patients at his own Memorial Hospital who had cancer and were near death. Even if the injected cancer thrived, their lives would be made no shorter.

Like most hospitals in the 1950s, Memorial provided little or no research oversight. It was up to Southam to decide what level of risk was acceptable, whom to recruit as subjects, and what—if anything— to tell them. It was up to him to balance the needs of future patients with the welfare of his experimental subjects. In his published articles, Southam never claimed that it was right for hundreds of subjects to be exposed to possible harm in order to save millions of lives in the future. But if that was his reasoning—if he believed that cancer's heavy toll justified experiments in which the subjects gave no consent because they didn't know they were in an experiment—he was not alone. Across the country, patients were being enrolled in a wide variety of studies, and thanked in print as volunteers, without ever knowing they had been part of an experiment.

Community doctors—local doctors with no involvement in research—were not much more forthcoming with their own patients. Practitioners varied, of course, but candor was not a prominent feature of the doctor-patient relationship of the time. Most, for example, did not tell their patients of a diagnosis of cancer, preferring euphemisms and evasions. But the community clinic and the experimental trial presented different moral contexts. Community doctors, no matter how little information they shared, were assumed to be working in their patients' best interests, whereas the loyalties of an experimentalist like Southam were at best divided. In his report of the experiment, Southam wrote that all subjects "were volunteers who were aware of the general purposes of the study and the nature of the implanted materials." This was simply untrue.

Patients at Memorial seem not to have objected, or even to have asked questions, when Southam told them he was going to test their immune systems. They would not have dreamed that the syringe in his hand was filled with live cancer cells. He injected a small amount—0.5 milliliters—of HeLa or another line of cancer cells just under the skin. As had happened with Senn, between five and ten days after transplantation, a nodule appeared, reaching a maximum diameter of about an inch.

Southam removed most of these nodules. Microscopic examination showed that they contained "healthy cancer cells." These transplanted cancers were under immune attack, and they usually disappeared in a few weeks. But in four cases, the cancer showed an alarming ability to grow and spread. Southam was able to remove the new growths in two patients. In two others, the transplanted cancer continued to invade, and in one, a woman with advanced uterine cancer, it spread to the lymph nodes in her armpit. The transplanted cancer would probably have killed her if her own malignancy had not done so first.

This experiment demonstrated that most of Southam's patients, even though they were seriously ill, retained sufficient immune competence to reject a foreign cancer. Would healthy people do so? Without exception? And, if so, would the rejection process differ in any way? For the next phase of his work, Southam asked Ralph Avis, the warden of the Ohio State Penitentiary, for permission to use prisoners as subjects. Avis agreed, on the condition that the scientist explain the experiment to them fully. Southam's ad in the prison newspaper asked for volunteers for the "injection of live cancer cells (taken from some person who has a cancer) into both forearms of the volunteer—by needle injection under the skin." There was no promise of payment or early parole.

Southam hoped for twenty-five subjects; he got 129. Many of the prisoners wanted to help in the fight against cancer; one had lost his father to the disease, another his sister. None had more reason to be fearful of cancer than an orderly in the prison hospital who nursed men dying of the disease—yet he, too, volunteered.

The prisoners were given their injections and soon developed painful lumps. They could not help being worried, but this lump was only a symptom of the energetic response of their healthy immune systems. Biopsies at Memorial had shown vigorous cancer cells, but when Southam biopsied the prisoners' lumps he found the foreign cells were under aggressive immune attack. Within a month, the injected cancers had vanished, the prisoners were relieved, and the results could be readied for publication.

All concerned were satisfied. The prisoners had endured some discomfort and taken some risk—otherwise there would be nothing to applaud—but they were unharmed. The *Reader's Digest* ran a laudatory article that quoted Warden Alvis: "this research has already paid off. It has provided all our men—the volunteers and the others alike—with a new sense of pride in our common humanity."

The NIH funded some of this work, and I suspect that Shannon was pleased—the experiment addressed an important problem, and it yielded both significant results and good publicity. Beecher, reading his journals in the Mass General operating room, probably also approved. The prisoners had given their informed consent, and the risk from Southam's injections was small.

With the prisoner research successfully completed, Southam's program entered a new phase. He had shown that prison inmates rejected transplanted cancer cells promptly, and that hospitalized cancer patients rejected the same cells more slowly or not at all. But why was the cancer patients' response impaired? Was it because of their malignancy? Or was there an immune defect that all patients with advanced disease would share—one that might also be seen, for instance, in Parkinson's disease? This question could be answered by another experiment, this time using patients with chronic nonmalignant diseases.

Southam's own hospital, Memorial, provided care for cancer patients only, but the perfect setting was in nearby Brooklyn: the Jewish Chronic Disease Hospital. Southam met with Emanuel Mandel, director of the department of medicine, in July 1963. In a follow-up letter, Southam reviewed the proposed experiment, complimented Mandel on the hospital's interest in teaching and research, and asked for permission to proceed with an investigation of the immune response of some of its patients. Mandel agreed.

There was still the nettlesome question of consent. Beecher's article had commented on its importance, but also left a loophole: When there is no risk, he believed, there is no need for consent. But who is to determine the level of risk? Southam considered his experiment to be almost risk-free, for by then he had injected cancer cells in hundreds of people and was confident that it was less dangerous than a routine spinal tap. In addition, he feared that informing his subjects that the injections contained cancer cells would provoke their "phobia and ignorance." And so he went to the chosen patients and, at their bedside, as he pushed in the syringe's plunger, he told them that this injection of "some cells" would "test their immune system."

Mandel was chief of the department, but every patient at the hospital had their own doctor. Many had not been consulted or even informed of the experiment, and when they heard about it, they were livid. Some of the patients had been refugees from the Third Reich; to their doctors, Southam and Mengele had too much in common. In the debate that erupted, Southam claimed that he had informed the patients about what he was doing and obtained their consent. His critics pointed out that one subject could speak only Yiddish, another had been deemed mentally unsound for years, and a third had advanced Parkinson's dementia. The hospital's board of directors held acrimonious meetings with heated charges and counter-charges.

A few years earlier, the *Reader's Digest* had lionized Southam and his inmate subjects. This time the story was different. The *New York Times* front-page headline read, "Hospital Accused on Cancer Study; Live Cells Given to Patients Without Their Consent, Director Tells Court." Senator Jacob Javits considered the study unethical and urged the NIH to require written consent from research subjects.

Southam himself was in no doubt that his work was highly ethical. In exchange for a small amount of discomfort and a risk that he considered to be practically zero, his subjects contributed to research that might one day lead to a vaccine for cancer. He felt the people who were upset simply didn't understand that any risks were dwarfed by the possibility of benefit to millions of people.

This utilitarian reasoning would have been persuasive two decades earlier, when James Shannon was supervising wartime research on malaria. As part of these investigations, scientists had been conducting experiments that were far more dangerous than Southam's. In one study, patients in the back wards at the Manteno (Illinois) State Hospital were deliberately injected with malaria, then given one of the drugs being evaluated to treat the disease. Typical symptoms included high fever, nausea, vomiting, and blackouts, with death a real possibility. These patients were chronic, back ward cases, and they were incapable of giving consent. But at a time when clerks and farm boys were being drafted and shipped to the Pacific, infecting the mentally ill with malaria was generally seen as asking no greater sacrifice of them than of everyone else. Nobody complained, major strides were made in the treatment of malaria, and Shannon received the Presidential Order of Merit.

When Beecher conducted the ether experiment on R.M., he assumed that the wartime assumptions still held—that the scientist, on behalf of the nation, could enroll someone in valuable research without their knowledge or consent. If, as a 19-year-old, R.M. had undergone a dilatation and curettage (D&C), probably the most common surgery for women of this age group, the time for the procedure would have been five to ten minutes, and the total anesthetic time about fifteen minutes, so Beecher's experiment may have tripled her time under anesthesia. It certainly increased her chance of anesthetic complications. This risk was acceptable, in Beecher's opinion. But public sentiment shifts with the times, and by the 1960s, public approval of involuntary sacrifice was long gone. It was replaced by a rising tide of concern for the individual and a new respect for the rights of groups like minorities, patients, women, prisoners, the disabled, and the mentally ill.

Southam was one of the many scientists whose indifference to the rights and welfare of their subjects had become a problem. But while the problem was everywhere, Southam's behavior was particularly egregious, as was his error in imagining the public's reaction. He imagined that people's fear of cancer would generate support and gratitude for his work; instead, it redoubled their horror at what he had done.

The Southam scandal of 1963 brought home to Beecher the futility of his impassioned article of four years prior. His proposal for reform had had, he admitted, "all the impact of a feather in a high wind," for the journals he read in the operating room showed a continued flood of abusive research. Most institutions showed scant interest in the problem, and scientists had a remarkable ability to rationalize their actions.

David Nathan was a good example. His moral standard was that he would not do anything to a subject that he would not be willing to undergo himself. This had a pleasing echo of the Golden Rule, but it would be fair to ask, what did this mean in practice? The answer, it turned out, was that Nathan was willing to draw his own blood, to sample his own bone marrow, and to inject himself with various substances, inadvertently giving himself hepatitis in the process. Did that give him the right to do the same to others, without asking for their consent? Nathan admitted frankly that this practice was wrong—fifty years later.

Back in the 1960s, Beecher recognized that his efforts at moral suasion had failed, but he remained deeply opposed to government oversight. As his next step in the campaign to reform research ethics and fend off regulation, he decided to issue a more forceful call for voluntary reform. The result would be the paper for which he is still celebrated today. He worked hard on successive drafts of this exposé, and by February 1966 it was within months of publication—and perhaps of shaking the medical establishment from its slumber. But time was running out.

SHANNON WAS OBSERVING THE PROBLEM of abusive experimentation as closely as Beecher. In fact, he had a better view, since his agency had funded some of the problematic studies, and it counselled scientists whose plans might endanger their subjects. In these efforts, the NIH had more influence than Beecher—it was, after all, writing the checks—but that influence was still limited. It was longstanding NIH policy not to tell scientists what they could or could not do once a grant had been awarded. NIH officials, many of whom had been productive scientists themselves, knew how often breakthroughs came from breaking free from the original research plan to pursue serendipitous opportunities.

Southam's experiments demonstrated how this laissez-faire policy permitted crises to blossom. After the first reports, the outrage grew only worse. Nathan Fink, a patient at the hospital, had received an injection, seen a lump develop and disappear, and been told that he had "good resistance." Some of the other subjects were demented, but Fink had all his wits about him. When he read about Southam's experiment in the newspaper, he realized that he was one of the subjects, and he sued. The hospital argued that liability should rest, not with it, but with the Public Health Service (of which NIH was a part), which had funded the research. The hospital demanded that the PHS take over the defense of the case. The government refused, and the controversy eventually played itself out. But the incident brought home to Shannon and his colleagues their legal and public relations exposure.

Southam was not the only NIH-funded investigator to give Shannon migraines, and to show the failure of Beecher's campaign of moral suasion. In 1959, Beecher had called on scientists to obtain consultation by a committee in case of doubt. Four years later, a Tulane surgeon transplanted chimpanzee kidneys to humans, all of whom died, without consultation of any kind and with no real hope of gaining useful information.

There was also growing international pressure for more formal subject protections. With the war over and the soldiers out of danger, Western societies developed a heightened concern for individual rights and liberties, including the welfare of research subjects. The World Medical Association debated a set of ethical principles for research with humans that became the Declaration of Helsinki. In the United Kingdom, the Medical Research Council, although it rejected any attempt at a detailed code, still issued a statement in defense of subjects' rights.

By September 1965, Shannon and other top officials were working to form a consensus in the PHS leadership about the necessity for federal action and the form it should take. They were struck by the success of the Clinical Research Committee, now in its thirteenth year of protecting the NIH's own hospital from scandals and lawsuits, and they decided to use the CRC as the model for a new system.

IN FEBRUARY 1966, NEW RULES mandating federal oversight of medical research were issued as Public Health Service Policy #129. The policy required each research institution to set up a committee to review proposed studies and ensure "an independent determination: (1) of the rights and welfare of the individual or individuals involved, (2) of the appropriateness of the methods used to secure informed consent, and (3) of the risks and potential medical benefits of the investigation." This was the first policy intended to protect subjects in federally funded research outside the walls of the NIH, the first attempt at a national replication of the CRC, and the beginning of today's IRB system.

This federal action ended Beecher's dream of voluntary reform. It had been a long shot from the beginning, since the chances were nil that an elite that was moving from one triumph to the next would admit that its achievements were morally compromised and limit its own professional freedom. Scientists did recognize that abusive experimentation could be a problem, but they tried to hide those abuses with euphemisms about volunteers and lies about consent. Beecher's only tool was moral suasion, and that was not enough.

The rationale behind the new policy was straightforward. Some research is malignant, and the new IRB system was intended to block it. Some research is benign—Fredric Coe's comes to mind—and the system should keep out of its way. And some research poses risk at a level that might be justified by the benefits to subjects or to society. In this case, an appropriate consent process is needed.

One of this system's great merits was its flexibility. It had been issued on the authority of the Surgeon General, and so, as time revealed problems in implementation—inevitable with any new program—it could be revised at the stroke of a pen.

The universities, medical schools, and other institutions that received federal funding already had committees to decide on matters of education, hiring, promotion, tenure, and much more. IRBs were created in the usual pattern, with committee members who were led by a powerful chair and supported by administrative staff, which could be as little as the part-time services of a single secretary.

Early reports of the new system's work were encouraging. Two years after its creation, the federal official who oversaw these early IRBs reported that proposals with "medical, psychological, or sociological" hazards had undergone a "dramatic" decline from 7.4 percent in 1966 to 1.7 percent in 1968.

The new system could be understood in ethical terms—it clearly implied that the welfare of subjects and of society are both morally relevant when designing an experiment—but it did not use the language of ethics. Shannon was no philosopher, and he and his colleagues saw no value in the development of an ethical code. He relied on the members of each institution's committee to make reasonable judgments, based on common sense and their own experience, and saw no reason to spell out the considerations they might take into account. He believed that Southam had been right when he injected consenting prisoner volunteers with cancer cells and wrong when he injected Nathan Fink and the other unconsenting subjects, and he believed that a committee of reasonable people needed no code of ethics to see the difference.

Beecher shared Shannon's disdain for codes of ethics. He opposed Shannon's regulatory solution, but he had nothing concrete to offer as an alternative. Nor was he willing to admit the impossibility of his goal of a purely voluntary system. He never stopped hoping that the ordinary humans who became scientists could develop an extraordinary ability to restrain their dreams in order to protect strangers from hazards that might never materialize.

Even after the IRB system was created, Beecher continued to campaign for his version of reform, which would have involved no regulation or mandatory review. In June 1966, he published the paper for which he is best remembered, "Ethics and Clinical Research." It does not mention Shannon's policy, issued the previous February, and refers only in passing to "so-called codes." Consent is important, he writes, but patients are easily persuaded to enter risky studies. "The more reliable safeguard" is therefore provided by "an intelligent, informed, conscientious, compassionate, responsible investigator." Such a scientist would never expose experimental subjects to harm for the sake of science.

Beecher's 1959 article had been long on generalizations, short on particulars, and bereft of impact. The draft of his new article cited fifty specific examples of "unethical or questionably ethical studies." He did not identify the studies by name, but scientists familiar with the field were in no doubt about which they were. One case involved an investigator who injected cancer cells into chronically ill patients; Beecher doubted that they had consented. Another scientist, studying hepatitis in mentally defective children, had obtained consent, but Beecher doubted that the parents were informed about the study's "appreciable hazards." In study after study, investigators claimed that their subjects had voluntarily accepted significant risk. Beecher scoffed at these assertions. There are, he granted, a few risk-tolerant people who might knowingly consent to almost anything, but most people would never risk health or life for the sake of medical progress. We can assume that when significant risk is present, genuine consent is absent.

These studies, Beecher wrote, were not the work of a few rogue investigators; they were done by highly-esteemed scientists pursuing research that their peers saw as important enough to justify the fiction of consent, even though the profession knew that these subjects were never adequately informed. The benefit to society implicitly justified the risk to which these subjects were exposed and the harm they suffered.

In October, 1965, the *Journal of the American Medical Association* rejected Beecher's paper. The following month, he submitted it to the *New England Journal of Medicine*, and in the final version he cut the fifty examples down to twenty-two. He knew that critics would argue that his examples, even if valid, were atypical and rare. He took a step toward countering that argument by explaining that these twenty-two studies were a selection from fifty that he had singled out. Those fifty contained references to an additional 186 studies that were also probably unethical, for a total of 236—surely an appalling amount of unethical research.

Still a loophole remained. Were even those 236 studies significant, when 137,000 medical articles would be published in 1966? Beecher solved that question elegantly. He had read through one hundred consecutive articles in an "excellent journal"; twelve seemed to involve unethical research practices. He concludes that "if only one quarter of them is truly unethical, this still indicates the existence of a serious situation."

Beecher did everything he could to ensure that his new broadside received widespread media attention. The first paragraph made the seriousness of his charge clear: "Evidence is at hand that many of the patients in the examples to follow never had the risk satisfactorily explained to them, and it seems obvious that further hundreds have not known that they were the subjects of an experiment although grave consequences have been suffered as a direct result of experiments described here." People in "sophisticated circles"—Beecher does not say just who—feared that attention to abusive research would halt progress. "But, according to Pope Pius XII, '. . . science is not the highest value.'"

The media response was very positive, generating stacks of letters and telegrams from a grateful public. Practicing doctors were appreciative as well; one wrote, "Thanks for your staunch defense of medical ethics. Most of us in practice are at a loss as to what is going on as regards experimentation; but we feel things have gone too far, and are not being guided by proper principles. I know you will get plenty of criticism from others, so I just want to commend you for your honesty and courage."

Many of Beecher's scientific colleagues saw things otherwise. He had never asked whether they believed their work was justified, and his apparent candor struck them as a sham. One wrote to the *Journal* to charge that he "quotes out of context, oversimplifies and otherwise distorts."

Beecher didn't care. He was done with private debate. Perhaps he did not have every detail right, but he had never claimed that every one of the studies was unethical. His point was that unethical research in medical experimentation had reached a crisis.

This impassioned essay is what Beecher is best remembered for today, so much so that some authors implicitly or explicitly credit its publication with forcing Shannon's hand in creating the IRB system. That exaggerates the article's influence, for although Shannon was aware of Beecher's campaign, Shannon had other reasons for addressing the problem. What's more, he issued Policy and Procedure #129 months before Beecher's exposé was published.

Beecher's achievement was not in inducing the government to act, nor did he seek government action—again, he was opposed to regulation. His contribution was to open up the closed world of medical research and bring the problem of unethical research to public attention, and for this he is justly praised.

Beecher did have to finesse some awkward problems in this essay, including the question of consent. In 1949, when he studied R.M.'s kidney function under anesthetic, it was not customary to obtain consent. His 1959 article justified not obtaining consent so long as the experiment is without risk. But this led to another problem. In his 1966 essay, Beecher pointed to the risks of anesthesia. In his own 1949 study of kidney function, R.M.'s anesthetic time was prolonged by forty-five minutes; in other subjects it was as much as seventy-four minutes longer, and the risk of complications increases with time under anesthesia.

He therefore faced a choice when crafting his exposé. He could have included this experiment among those he publicly condemned, a candid admission that his own experimental work once fell below the high standards he now promoted. But seven of the "unethical or questionably ethical" studies on his list were more than a decade old, those scientists' understanding of how subjects should be treated might have changed just like his own, and he had no interest in analyzing the shift in acceptable morality over time. Instead, his article implicitly argued that certain principles are absolute, including the one he kept repeating: that the public good never justifies placing an individual at risk. He did not mention his own problematic research.

In his call for higher ethical standards, Beecher faced a second, even more vexing problem: Who should determine how general standards should be applied to the specific facts of any given case? His 1959 paper had supported committee review, or, in his words, "group decision supported by a proper consultative body," and he had cited with approval the CRC's review process. But several of the unethical experiments on Beecher's list had been conducted at the NIH and might well have been reviewed by the CRC. One study conducted outside of the NIH, on antibiotic side effects, had been approved by the local hospital's review committee. And Krugman's hepatitis protocols had been reviewed and approved by four separate committees at the medical school, state, and federal levels.

In this new paper, Beecher could have argued that committee review had failed, and perhaps sought to show how it could be improved. He could also have proposed another oversight method. Instead, he omitted any discussion of committee review, endorsing in its place an investigator who is "intelligent, informed, conscientious, compassionate, [and] responsible." This solution fails to come to terms with the basic problem: Most scientists, including those whom Beecher attacked, already saw themselves as intelligent, informed, conscientious, and so on. An oversight committee could provide an objective perspective, but Beecher was evidently through with committee review. He trusted only his own judgment. This left other scientists in the dark as to how they could do work that would meet Beecher's approval. He had identified an urgent problem, but he had no real solution to propose.

WITH BEECHER HAVING BROUGHT PUBLIC ATTENTION to research ethics as an important and evolving area, the American Academy of Arts and Sciences decided to devote its considerable prestige and resources to an in-depth look. In 1967 and 1968, the AAAS held conferences of influential scholars and scientists to discuss the problem. The meetings revealed some bitter divisions, particularly with regard to Krugman's work at Willowbrook.

Beecher had set the stage with his 1966 essay. Its brief summary of the Willowbrook experiment noted that "artificial induction of hepatitis was carried out in an institution for mentally defective children in which a mild form of hepatitis was endemic." The ethics could not be more straightforward: "There is no right to risk an injury to one person for the benefit of others." The parents did give consent, but Beecher doubted that they knew how risky the experiment was. Privately, Beecher wrote to a critic, "The thought that some would have agreed that deliberate infection was all right since the subjects were mental defectives gives me the Nazi shudders."

Art Caplan is one of the modern ethicists who agree with Beecher that the studies were wrong, no matter how mild the illness in the children and how great the benefit to society. Scientist and vaccination pioneer Maurice Hilleman concurs, denouncing the Willowbrook studies as "the most unethical medical experiments ever performed in children in the United States."

But there is no consensus now, and there was no consensus at the AAAS meeting. In his commentary, Louis Lasagna, a professor at Johns Hopkins, considered Beecher's analysis too simplistic. He agreed that the work "superficially seems disturbing," but argued that deeper study reveals the experiment's moral strengths: Infection was almost inevitable, the children were protected from the other diseases that were rampant in the institution, the protocol was approved by the relevant agencies, and the parents consented. After weighing all the relevant considerations, he concluded that the study was unobjectionable.

Having defended Krugman, Lasagna then moved on to attack the "tendency for people to moralize about the ethical problems in human research in terms of black-and-white categorical imperatives. ... Absolutist doctrines seem no more defensible in this area than in others." This was an obvious reference to Beecher, who was prone to absolutist pronouncements, and Lasagna did not spare his colleague. "What sometimes passes for ethical profundity may, in fact, be only shallowness and an irresponsible or arrogant failure to appreciate the richness of the moral alternatives and the subtlety of the ethical issues."

Krugman had already been defended by Walter Goodman, a reporter and critic for the *New York Times*. Writing in 1967, Goodman asked whether the Willowbrook study was justified by the dramatic reduction of hepatitis at Willowbrook, and quoted Beecher's absolute denial: "There is no right to risk an injury to one person for the benefit of others." Goodman, unpersuaded, asked, "Even a million others? Even an enormous benefit? Even an infinitesimal risk of an inconsequential injury?"

We know now that "even a million others" underestimated the eventual value of the research. It was foundational to the development of vaccines against hepatitis A and B, which have saved many millions of people from these sometimes-fatal infections. The polio vaccines got bigger headlines, but the hepatitis vaccines may have saved more lives.

Krugman's work was also defended by the Willowbrook parents' association, which was well aware of the controversy. One year after Beecher's essay was published, the parents' group, which fought to improve the institution's funding and operations, gave Krugman a plaque that praised his "distinguished, pioneering, humanitarian research in the prevention of infectious diseases and their resultant complications in children, born and unborn."

I've talked with many people about the Willowbrook experiments, and most have a visceral reaction for or against the study. Those who feel the research was wrong tend to be revolted at the contaminated milkshakes and empathize with the children. Those who feel the research was morally justified are sometimes influenced by the lives it has saved, but they are also swayed by the suffering of the nurses—people who chose to care for the children and became terribly sick as a result. Beecher and Krugman came to the issue with their own professional experiences. Krugman, as a pediatrician, was perhaps influenced by the mild illness he observed in children at Willowbrook and elsewhere, while Beecher regularly saw the fulminant disease that swept adults to their death.

The AAAS conference organizers knew that the Willowbrook controversy would be on the agenda, and they expected several participants to comment on it. The roster included several doctors, along with lawyers, sociologists, and others. But there was a lacuna that could prove awkward: Although many of the attendees had strong opinions about the ethics of research, not one had formal training in philosophy. There was a real danger that this inquiry into the ethics of human experimentation would need to proceed without the participation of an actual ethicist.

There wasn't even an obvious candidate for the role. No living philosopher had given sustained attention to the challenges of the field of research ethics; it was virgin philosophical ground, and its value to an academic's career was unproven. So the working group no doubt considered itself fortunate to persuade a New York philosopher to join the conference and contribute a paper to the proceedings. With the scientists at ethical loggerheads, into the fray stepped Hans Jonas, a nattily dressed philosopher in his mid-sixties with a habitual cigarette in one hand.

Jonas was born in the German town of Mönchengladbach and obtained his PhD for a study of Gnosticism in 1928. His supervisor was Martin Heidegger, author of *Being and Time*, a landmark achievement in existential philosophy. Their productive collaboration was ruptured when Heidegger joined the Nazi party and signed a public oath of loyalty to Hitler. Jonas watched during the 1930s as, step by step, German Jews were stripped of their rights, their property, and their lives. The last straw was when the German Association for the Blind expelled its Jewish members. Jonas was not a member (nor was he blind), but he was Jewish, and he decided it was time to flee.

When war came, Jonas was living in the British protectorate of Palestine. The British army, finding a German-trained PhD among their recruits, invited him to work in military intelligence—a prospect surely more interesting, and much safer, than that of the ordinary soldier. Jonas refused: He wanted to fight the Nazis in person, with a rifle, and so he joined the Jewish Brigade of the British Eighth Army. After the war, he learned that his mother had been murdered at Auschwitz.

Jonas eventually made his way to the New School for Social Research in New York. The New School had a long commitment to academic freedom, and it welcomed scholars who had fled fascist Europe, like Jonas, Hannah Arendt, and Erich Fromm. It was in this new home that Jonas developed a reputation as that rare animal, a philosopher who writes clearly about practical matters, and it was here that the conference organizers contacted him. Would he be interested in commenting "philosophically," first in a talk at the conference, and then in a published paper, on human experimentation?

As Jonas considered their request, he found an emergent field that was largely devoid of ethical theory. As the only philosopher at the conference, he was thus in a unique position to shape future thinking about abusive research. In his subsequent essay, Jonas mentioned no specific experiment, but every word revealed his revulsion to the studies in Beecher's article and to the atrocities of the Nazi doctors. The most notorious of those doctors was Josef Mengele, who conducted his experiments at Auschwitz, where Jonas's mother had died. She might even have been one of his victims. In Mengeles's experiments, Jonas saw a tragic conflict between the all-powerful Nazi state and the defenseless individual. He saw this conflict recapitulated in experimentation in the United States, and his essay emphasized the absolute necessity of protecting the research subject.

American scientists claimed that their research was made moral by the consent of their subjects. Beecher had already asserted that subjects were often duped and their consent a sham. Jonas went further, arguing that subjects almost never understand research well enough for their consent to be morally valid. Further, no matter how polite the invitation to enroll, consent is inevitably extracted under pressure. "The mere issuing of the appeal, the calling for volunteers, with the moral and social pressures it inevitably generates, amounts even under the most meticulous rules of consent to a sort of *conscripting*."

In his most compelling passage, Jonas attacked the belief that we must pursue cures for the diseases that ravage us, that we cannot afford to forego continued medical advances. To the contrary, he wrote, we must accept what we cannot avoid, and that includes disease, suffering, and death. What society genuinely cannot afford is "a single miscarriage of justice, a single inequity in the dispensation of its laws, the violation of the rights of even the tiniest minority, because these undermine the moral basis on which society's existence rests." He concluded that "progress is an optional goal."

Beecher had believed that experiments he considered "innocuous" required no consent at all. Jonas argued that there is no such thing as an innocuous experiment, since research subjects who are not in physical peril are nonetheless stripped of agency. "What is wrong with making a person an experimental subject is not so much that we make him thereby a means (which happens in social contexts of all kinds), as that we make him a thing—a passive thing merely to be acted on. ... This is different from even the most exploitative situations of social life."

Beecher brought public attention to the ways in which research can harm its subjects. Jonas went further, to argue that abusive research devastates the moral quality of the nation. Allowing any compromise of the individual's rights and welfare would be to take "imperceptible steps" that could lead to "threatening Utopias on our own horizon." Progress, however desirable, will never justify compromising the values that we as a society cherish.

Jonas's essay also explored the moral challenges that societies face when they confront epidemics and wars. In circumstances like these, a society's continued existence may require that people be conscripted into experiments. He also discussed the impact of high or low birth rates, of diseases that affect the young or the elderly, and more. The essay offered a nuanced analysis. But too often it is remembered only for its fundamental axiom, that research subjects must be resolutely protected, and for its corollary, that progress is optional.

Time would show the limitations of Jonas's approach. If we adopted his exacting standards for the absence of risk, the perfection of consent, and the protection of research subjects' agency, scientists might as well close up shop. This is to put it in practical terms, but there is a moral problem with Jonas's view as well. By focusing on abusive research, Jonas misses the moral value of experiments that are properly conducted. By "moral value" I am referring to, for instance, the value in saving patients with kidney stones from agony, surgery, and narcotic addiction. Medical breakthroughs help people—sometimes in great numbers—but Jonas took this reduction in suffering and death, pinned it with the abstract label "progress," and asserted that it was optional. In contrast, the IRB system launched by Shannon was intended to consider both the need of the subject for protection and the need of society for progress, and thus to balance these considerations.

The IRB system itself began to change and grow. In its early years, federal officials gradually expanded the scope of what IRBs should review and required increasingly detailed records. As their responsibilities expanded, IRB chairs needed more administrative help from people with greater skills. Chairs were happy to hire administrators who could master the regulatory details—who could learn, for instance, the eight required elements of informed consent as well as the circumstances in which exceptions were permitted, and keep an eye on protocols to keep the committee from error. Scientists with questions, who would once have spoken with the chair, now found their inquiries answered by full-time IRB staff.

SHANNON AND HIS ADVISORS HOPED that the new system of review would make abuses in PHS-funded research a thing of the past. None appeared to have considered the possibility that the most monstrous experiment was being conducted by the PHS itself. This was the Tuskegee syphilis study, the best-known and most devastating abusive research ever conducted on American soil. Its lesson—that the vulnerable must be protected from abuse—should never be forgotten. But, as often happens with scandals, the government's reaction led to an unanticipated consequence: today's dysfunctional IRB system. But before we get to that dysfunction, we need to consider the syphilis study itself.

# 3

# Congress Steps In

PETER BUXTUN WALKED PURPOSEFULLY down tenement hallways scented by uncollected trash and urine, all the way to the last door, the one with four locks, where you had to bang two or three times before anyone answered. The voice that answered was never welcoming.

For Buxtun, who sported shaggy hair and a full beard, it was part of the job. He wasn't a landlord or a process server or a bill collector; he was a minor government functionary who believed deeply in his official mission. His employer was the Public Health Service; his enemy was a disease that had afflicted humanity for 600 years. It was 1966, so San Francisco's Summer of Love was still a year in the future, but the summer of sex was eternal. Buxtun tracked down people who had gone to a public clinic complaining of a vaginal discharge or a penile sore or a peculiar rash. He had their test results.

Once Buxtun talked his way inside—he was good at that—he would explain that the tests showed syphilis (the clinic tested for other diseases, but syphilis was the only one that routinely killed). By the time of Buxtun's visit, the discharge, sore, or rash had often

68

disappeared, so his challenge was to persuade his clients that, even though their symptoms were gone, they still carried a dangerous infection. They might have no further problems, but they might also develop fatal brain or heart complications. What's more, a man with latent syphilis can infect his wife or girlfriend, and her infected baby can suffer from anemia, seizures, and birth defects. Buxtun showered the skeptical with graphic details of the ravages of untreated syphilis, which could, among other horrors, so enlarge the aorta as to create a fist-sized bulge in the front of the chest.

That was the bad news. By this point, some of Buxtun's customers were ready to hear the good news and eager to get treated. Thirty years earlier, state-of-the-art treatment of syphilis had required seventy weeks of injections with guaranteed toxic side effects and the mere possibility of a cure. But that dismal picture had changed dramatically when penicillin became widely available after World War II. By 1966, treatment for syphilis had become safer, simpler, and much more effective.

Most of Buxtun's customers agreed to return to the clinic for their shot, and to give him contact information for their sexual partners. That way he could climb more stairs and walk down more dimly-lit hallways and do it all over again.

Buxtun was bearded in the style of the 1960s, but he was no hippie. Historian Susan Reverby describes him as "a libertarian Republican, former army medic, gun collector, and NRA member with a bachelor's degree and some graduate work in German history." He was also, like Hans Jonas, a refugee; his family had fled Czechoslovakia and the Nazis when he was an infant. He was proud to be part of the PHS's nationwide campaign against syphilis.

His feelings about the agency were about to change dramatically.

In the fall of 1966, Buxtun heard an older PHS employee talking about a study in which government doctors *prevented* men with syphilis from getting treatment. Finding this hard to believe, Buxtun wrote to the study coordinators for more information. They sent him a thick packet of medical journal reprints, and he found to his amazement that it was true. In 1932, PHS doctors had identified a large group of infected African American men living in the Tuskegee area and had spent the succeeding decades studying their deterioration and actively keeping them from getting care. The men had no idea that they were in an experiment or that they were being denied care. While Buxtun was begging syphilitics in San Francisco to get their shots, his employer had decided that a large group of men in Alabama should be permanently denied treatment. Responsibility for managing the study had originated in the PHS and was eventually transferred to the Centers for Disease Control.

In 1966, Buxtun wrote William Brown, a senior CDC official, to attack the study's ethics. Brown responded that this was a low-risk, voluntary study. Buxtun wrote Brown again in 1968, at a time of widespread urban rioting, not long after Martin Luther King, Jr. was assassinated, saying, "The group is 100% negro. This in itself is political dynamite ..."

Brown could dismiss Buxtun's concerns about the study's ethics, but perhaps the politics deserved a closer look, and so the CDC convened a panel to reevaluate the study in 1969.

The panel seemed incapable of seeing the study as racist, but it did recognize that depriving hundreds of Black men of treatment for a fatal disease might still be bad optics. One of the (all-white) panel members commented, "This is not a Study that would be repeated now. The public conscience would not accept it." But after spending three hours considering the study's methods, findings, and prospects, and thinking through the pros and cons of terminating the study, the panel chose to continue it until the last subject died.

Buxtun, who viewed the Tuskegee study as morally equivalent to the Nazi atrocities, was entirely unpersuaded. Ever after he left the PHS, he would tell friends about the shocking experiment. One evening in 1972, over dinner, he told the story to a group that included a reporter. That is how the Tuskegee syphilis study came to America's front pages.

The study was not born evil. It began in 1930 as a joint project of the PHS and the Julius Rosenwald Fund to identify *and treat* people with syphilis in the Tuskegee area, using the then-standard therapies of arsenic and bismuth. This treatment phase was cut short when the Depression choked off the study's funding. In 1932, instead of letting the project lapse, PHS scientists decided to compare the health of approximately 400 infected men with 200 healthy controls, work that they believed they could complete within six to eight months. Then they changed their plans again, and decided to follow the men indefinitely with periodic tests, medical examinations, and, ultimately, autopsies.

The investigators immediately began lying. They gave the men aspirin and vitamins and claimed that the examinations and blood tests were also part of their treatment. But treatment is exactly what the men did not get; the scientists worked to make sure that no doctor, public or private, provided appropriate treatment, even after penicillin became available.

By the time the study was shut down, 128 of the men had died of syphilis or related complications, many of their wives had become infected, and some of these women bore infected children. The Tuskegee Institute provided some facilities for the project, and PHS called the experiment the Tuskegee Study of Untreated Syphilis in the Negro Male. This usage rankles Tuskegee University, which argues that it should be called the United States Public Health Service Syphilis Study. The school has a point! But the battle has been lost in IRB circles, where it is usually referred to by the place name alone, as in, "We should reject this proposal—it's like another Tuskegee."

The exposure of the Tuskegee study in 1972 generated shock waves across the nation and around the world. Congress was moved to action, and it attempted to ensure that such a horrific abuse of human rights in the name of medical science would never again be permitted. Today we are still grappling with the aftershocks of the study, and of Congress's response.

UNDER SHANNON'S OVERSIGHT, the NIH's Clinical Research Committee had balanced participant protection with the societal benefits of research and the right of prospective subjects to accept a measure of experimental risk and burden. That balance was gradually lost after the system was expanded nationwide in 1966.

Those changes had begun even before the exposure of the Tuskegee study. In the early years of Shannon's system, IRBs could function like any other university or medical school committee, with routine oversight by higher institutional officials. But in 1971, the Department of Health, Education, and Welfare revised the guidelines to deny scientists the right to appeal to the institution's leadership. Their only recourse was to appeal to the same IRB, or to a similarly-constituted committee. It became illegal for a higher official, such as the institution's president or its dean for research, to intervene. An IRB that said "No" operated, in effect, under a presumption of infallibility. That remains the case today.

The infallibility clause was added quietly, with little or no comment. Evidently this provision was viewed as so sensible that it needed no justification. As one IRB manual explains, "the rights and welfare of human subjects must always take precedence over the needs of science or fiscal considerations. ... It is important to preserve the autonomy of the IRB and insulate it from pressure exerted by individuals or groups with special interests that are inconsistent with the mission of the IRB." The committee, in short, "must be able to act as an independent and objective body without answering to multiple masters who may have different agendas."

I know of no published data in support of this belief that the IRB must be protected from interference by a morally compromised university president or medical school dean. Shannon never forced the NIH's own CRC to approve dangerous research, and I am unaware of this problem arising with any other IRB before the 1971 rule change. IRB infallibility would still be sound policy if IRBs were wiser than presidents. But there is no reason to believe that this is so, and while presidents are fallible, so too are IRBs, which may have political agendas, be vulnerable to error, and serve as the instruments for jealousy and academic infighting.

This change in the management of IRBs was just the first step in the gradual movement of the system toward dangerous imbalance. That change was accelerated—understandably, but regrettably—in the wake of the Tuskegee revelations.

Congress, which felt it must respond to the Tuskegee outrage, passed a law in 1974 known as the National Research Act. The act was a good-faith effort to prevent further scandals. But it was this act, and the mindset that accompanied it, that drove the IRB system to its present condition.

Recall that Policy and Procedure #129, which created the IRB system, had been issued on authority of the Surgeon General. The system could be improved in minor ways, or rewritten from the ground up, by a stroke of an official pen. The rule that the dean of a medical school could not tell an IRB that its requirements were unreasonable, for instance, could have been overturned. This flexibility was an invaluable asset for a system so young, charged with oversight of an industry so vast. The National Research Act ended this flexibility and grounded IRB oversight in the cement of federal law. This is why meaningful reform will require Congressional action.

Reform will also require change in the mindset of the people who oversee IRB operations. The first layer of oversight is provided by the Office for Human Research Protections (formerly known as the Office for the Protection from Research Risks) and the Food and Drug Administration, but these agencies are not the final word. They work under the supervision of Congress, which can increase their funding if it feels the money would help their work. Congress also has other options that are less appealing to the agencies.

You might suspect that the power of Congress to impose its will on thousands of IRBs is purely theoretical. It is not. In fact, it is a practical reality, and it was Congressional pressure that ushered in the most tumultuous period in IRB history, a period whose effects, in terms of IRB rigidity and dysfunction, remain manifest today.

GARY ELLIS DIDN'T SEE HIMSELF AS A FAILURE, but as they listened to Ellis describe his work, Representatives Edolphus Towns and Thomas Barrett were coming to that view. Congressional hearings can be that way.

It was the summer of 1998, and a subcommittee of the Committee on Government Reform and Oversight was meeting to investigate the performance of the IRB system. The Inspector General of the Department of Health and Human Services had issued a report suggesting that IRBs were too overworked to serve as reliable protectors of human subjects. The report called for reform, and Towns and Barrett were among the representatives who were looking for answers about this apparently-overwhelmed system

Ellis had led OHRP's predecessor, the Office for the Protection from Research Risks, for five years. During that time, OPRR reviewed new IRB applications, educated underperforming IRBs, and responded to questions and complaints. Ellis believed that his agency was a vital safeguard against scientific misconduct. The danger of abusive research was real, but Ellis believed that OPRR was doing a good job of containing it.

Still, Ellis agreed that reform was needed. He believed that research institutions should be required to give their IRBs more staff, space, and money, so that they could conduct more thorough reviews. Such reforms would also create and preserve documentary proof of those reviews in case an OPRR inspector came to call.

Representative Towns wasn't interested in documentation. He suspected that OPRR was asleep on the job, and he suggested that the system "needs to be torn down, rebuilt from scratch. Reforming is not enough." He began with a basic question: "How many IRBs do we have in the United States?" A federal official who was testifying with Ellis ventured that there were at least 3,000 IRBs, and perhaps as many as 5,000, but that "nobody knows for sure." Ellis was forced to concede that he also had no idea how many IRBs were, in theory, under his diligent supervision. Presumably his office maintained a file for every active IRB, but nobody had thought to count them.

Towns snapped, "Now, you can understand why I don't like the word 'reform.'"

Beneath this barbed exchange lay an unspoken question: What was OPRR supposed to be doing? Shannon's goal for the IRB system was to balance the twin needs of research subjects for protection and of society for the fruits of research. Representatives Towns and Barrett remembered Tuskegee, and they apparently assumed that medical research was fraught with risk. The job of the IRB was therefore to police untrustworthy scientists to reduce that risk, and the job of OPRR was to punish IRBs that failed in their duty.

The attack appeared to catch Ellis flat footed, and perhaps he felt that it would be unwise to disagree with the panelists. Yet he could have rebutted their concerns had he chosen to do so.

His best defense would have been to point out how seriously the public interest would be harmed if the agency focused only on reducing risk. In 1996, just two years earlier, the General Accounting Office, in a report to the Senate, recognized that "the government and the research community, whose ultimate goal is the advancement of scientific knowledge, struggle to balance two sometimes competing objectives—the need to protect research subjects from avoidable harm and the desire to minimize regulatory burden on research institutions and their individual scientists." The GAO noted the impossibility of perfection: "No practical level of oversight can guarantee that each researcher will protect subjects with complete integrity." And it gave the system credit: "The detection of instances of potential or actual harm to subjects both demonstrates that abuses can occur and suggests that the current oversight activities are working."

Ellis could also have argued that IRBs were doing a great job at identifying abusive protocols and blocking them. If scientists had learned that abusive protocols would be rejected, and IRBs blocked the few potentially abusive protocols that were nonetheless submitted, OPRR would have little occasion to impose sanctions.

But while this seemed a likely explanation, neither Ellis nor any other federal official had ever asked IRBs to submit data on their work. How many protocols did they review in the course of a year? Of what kind? Involving how many subjects? Nobody knew. Nor had Ellis asked IRBs to provide details about how many protocols they rejected as potentially abusive, and the reasons for those rejections. (Today it is easy to count the number of IRBs on an OHRP website, but the agency still requires no systematic reporting of their activity.) In 1998, this cost Ellis a chance to show the good work that IRBs, and OPRR itself, were doing. Of course, the results might also show that scientists had learned their lesson, and knew better than to propose the kind of experiments that had caused so much trouble decades before. In either case, subjects were being shielded from unnecessary harm.

Instead of defending his agency, Ellis agreed with Towns that OPRR's role was to protect subjects, leaving the importance of medical progress out of the discussion. The fact that the scientists might do something useful, even important, seems not to have entered his head.

Years later, he told a story that was a shocking confirmation of how little he cared about science. As OPRR director, he had been consulted when a political scientist and his IRB were at loggerheads. The scientist wanted to conduct research using data about elections and public opinion that was widely available and contained no individual identifying information. This would be about as low risk as research can get. The regulations recognized this and stated quite clearly that work of this kind was altogether exempt from review. The political scientist was appalled to discover that his IRB refused to acknowledge this exemption. When a university official called OPRR to get a definitive opinion, Ellis backed the IRB.

As Ellis explained, "The only thing that I heard was the institution's calling, asking me for an official reading on whether it's exempt. And it's not ever going to be exempt, if you call me on the phone and ask me that." He explained his reasoning. "At some level," he continued, "I've taken care of the human subject. That's the highest priority. The second priority is self-preservation of the bureaucrat. And the third priority is the researcher."

Return now to 1998 and Washington, DC, where the hearings were threatening Ellis's second priority. Representative Barrett asked Ellis what sanctions he could impose on hospitals, universities, or medical schools that violated the regulations. Ellis replied that while he could not punish them for past misdeeds, he could suspend their authority to conduct federally-funded research, a step that "is viewed by the institutions as a death penalty. It's extremely serious. The Federal research dollars stop flowing if we make that judgment." Barrett retorted, "I understand that but it seems that your organization is far less likely to give a death penalty. I just don't see you doing it." Ellis was forced to concede that, in his years in office, he had never taken such a drastic step.

Christopher Shays, the chair of the subcommittee, called OPRR's enforcement efforts "pathetic." Towns had already suggested that the ultimate punishment might better be meted out to OPRR itself. "I really have a great difficulty here with the whole thing about reform. We have so many problems here, we have to destroy this one and start over."

THIS CONGRESSIONAL PRESSURE MOTIVATED ELLIS to take the hammer off the shelf, and over the next three years his agency temporarily shut down some or all federally-funded research at a dozen institutions. At the University of Pennsylvania, a teenager with a rare disorder had died in an early attempt at gene transfer. At the other institutions that were sanctioned, however, there were no other deaths and few serious problems.

This paucity of injury and death was not surprising. Robert Levine, one of the IRB system's architects, had reviewed the evidence from institutions of every kind and concluded that "the role of research subject is not particularly hazardous in general." Even the teenager's gene transfer death was the result not of callousness or scientific adventurism (the protocol had undergone intensive scrutiny before it was launched) but of an immunological reaction that had never previously been observed.

In the absence of actual abusive research, Ellis reached for any handy excuse. OPRR's primary targets were IRBs that had failed to follow what the agency considered proper procedures, or had failed to document their compliance. The agency sanctioned Duke, for instance, because its IRB had failed to provide written justification for its decisions to approve research and uncertainty about whether a quorum was present at every meeting.

At Virginia Commonwealth University there were no injuries, but there was at least a complaint from a member of the public. A genetics researcher there had mailed out a family history questionnaire to subjects in a study of twins, after obtaining their consent. The father of one of the twins in the study complained to OPRR: He had not given his consent, and he felt his privacy was being invaded. This case involved complicated issues regarding who has a right to share what information with researchers—issues that are still being debated today—but there was nothing complicated about OPRR's response. It faulted the IRB and shut down the study until it could be reviewed again and the problem solved. This by itself was not obviously unfair.

But OPRR also shut down more than a thousand other studies with little or no connection to genetics, apparently on the theory that this was an IRB that could not be trusted. The university created an emergency task force of senior administrators, chaired by the president, that met every other day. Every one of the suspended protocols had to be re-reviewed by an independent IRB—to be paid by the university—before they could resume.

Other institutions were punished for what they considered trivial reasons. In one case, OHRP was concerned that IRB members who were initially present might leave early or even step out of the room for a minute. The agency required that committee to document that a quorum was present for every one of the hundred or more items on the agenda. The FDA, which joined in this disciplinary surge, was equally concerned with attendance. Its investigators discovered one IRB that defined a quorum as "more than one-half." The agency threatened to issue a formal reprimand because, in its view, a quorum should be "one-half plus one."

At every institution that was punished, as one observer wrote, "Participants cannot receive treatments, enroll, or be recruited; results from time-sensitive studies cannot be reported; and data cannot be analyzed. Suspension means that there is no money to pay graduate students, travel to conferences, or purchase equipment. It means researchers may lose months, if not years, of work." The Chancellor of the University of Illinois at Chicago, who was responsible for an institution with a budget in the billions, was fired after OPRR faulted the institution's processes for approving informed consent.

The rest of the research world watched with growing horror. When institutional leaders saw their peers sanctioned for trivial offenses, they knew they could be next, and it made them crazy. In 2000, the *Chronicle of Higher Education* commented, "Across the country, university administrators and researchers are worried, even panicked, that the same thing could happen at their institutions, with millions of dollars of research funds ... at stake." Now IRBs, "terrified at the specter of their institutions losing all federal funds, opted overwhelmingly for conservatism." They could do no less, since they were now "taking on the unspoken role of protecting the institution from both lawsuits and the suspension of federal funding." This leads IRBs, in the words of legal scholar C. Kristina Gunsalus, to make "overzealous demands that impede research and discredit the IRB."

FEDERAL PRESSURE TRANSFORMED IRBS NATIONWIDE, a process that is beautifully described by sociologist Sarah Babb, who interviewed dozens of administrators and officials. Before 1998, institutional leadership was satisfied when the IRB chair (who might be volunteering their time) made decisions that seemed reasonable and avoided scandals and subject injuries. The federal crackdown made it clear that more was demanded than any chair could provide. The chair, after all, was usually a busy faculty member with classes to teach, patients to see, and research of their own to conduct.

The post-1998 changes reshaped how IRBs operated and how IRB oversight was experienced by scientists. Institutions hired more staff to make sure that all the records that federal officials might demand were pristine. This additional staff needed higher levels of education and training, since they were given much of the responsibility for ensuring that the regulations were followed punctiliously. IRBs had once been paper-based and responsive to investigators' individual circumstances. The new systems were software-based and required every investigator to answer each question on the screen before proceeding to the next, whether or not the issue had any relevance to this particular study. Investigators with questions about any of this might find it difficult to get an answer from the overtaxed IRB, and any response was now likely to come from one of the staff rather than the chair. All of this activity was more or less invisible to IRB members (except for their own experiences as investigators), and the fraction of decisions that were made outside of formal IRB meetings increased.

Federal pressure gave IRBs, acting to protect their institutions, greater motivation to force investigators to comply with every demand, which is why Fredric Coe could not persuade his IRB that compiling a new bibliography with every new study served no purpose. In this new system, the IRB staff assumed much of the responsibility for day-to-day decision making, and the IRB chair now shared power with the chief administrator. IRB members continued to review protocols and vote in meetings, and many continued to believe that the committee's work was essentially ethical, even though it had become dominated, outside of the meetings, by increasingly detailed bureaucratic requirements.

The increases in staff to accomplish this were substantial. The staff of the Northwestern IRB, for instance, grew between the late 1990s and 2007 from two people to forty-five. These fortified IRBs were in no doubt that their mission now extended beyond protecting research subjects. As Northwestern's Caroline Bledsoe notes, "the IRB's over-riding goal is clear: to avoid the enormous risk to the institution of being found in noncompliance by OHRP."

It might seem that sufficient diligence in all of these areas would ensure success, defined here as avoiding federal sanctions, but it did not. As Babb writes, because OHRP's interpretations of the regulations "were ambiguous and inconsistent, there was still widespread confusion about what compliance entailed. This caused IRB offices to go above and beyond the rules, just in case." Institutions reorganized IRBs so that operational staff, who were responsible for complying with federal regulations and guidance, were themselves subject to institutional oversight. "The production of auditable hypercompliance was labor-intensive and created multiple levels of obstruction to the research process." There were problems with IRB review long before the death penalty era, but they were now far worse. Scientist dissatisfaction increased.

The result was unhappy for Ellis as well. He had previously been savaged by Congress for supposedly doing too little; now he had done too much. In 2000, the Department of Health and Human Services leadership moved OPRR up in the department's hierarchy and gave it a new name; it would henceforth be called the Office for Human Research Protections (OHRP). This refurbished agency would need a chief, and Ellis applied for the job. The search criteria, however, specified that a successful candidate must have a distinguished research background and be "viewed as a statesman." Ellis was soon gone.

But Ellis had never been the real problem. He was only following the belief, conspicuously shared by members of Congress, that the system should act as a police force, focused on keeping reckless scientists from harming their experimental subjects at practically any cost.

Every institution knew that they could be next. Greg Koski, who had been IRB chair at Massachusetts General Hospital, was called one day by Mark Yessian, the federal official who had written reports that were critical of the IRB system. Yessian wanted to chat about the protection of human subjects. As Koski later recalled, "it scared me to death."

Koski's story had a happy ending, for instead of his institution being shut down, he was offered, and accepted, the job of first director of OHRP. But Koski was well aware of the harm caused by the research suspensions that Ellis had begun. He later wrote that "the suspensions created a crisis of confidence and a climate of fear, often resulting in inappropriately cautious interpretations and practices that have unnecessarily impeded research without enhancing protections for the participants. Such 'reactive hyperprotectionism' does not usefully serve the research community, the participants or the public, and it should be avoided."

Unfortunately, reactive hyperprotectionism has not been avoided. It has grown, and today it dominates IRB thought and action. Let's turn now to the ways in which this harms all of us, beginning with IRBs' management of risk.

# 4

## Risks and Benefits,
## Real and Imaginary

R OB KNIGHT IS AN EXPERT in the human microbiome—the bacteria, fungi, and other microbes that live upon and within us. For most of the twentieth century, few of these germs had been identified. Toward the century's end, new experimental methods revealed an astonishing abundance and diversity of microbial life living on our skin, with other and quite different communities in our mouths, vaginas, and guts. The microbes in our gut alone weigh as much as five pounds and are ten times as numerous as all the cells in our bodies.

Our understanding of these microbes' function has been transformed as well. In the older conception, which held sway until late in the twentieth century, these germs were assumed to be either bad guys that threatened disease or harmless vagrants who pursued their own interests. Modern genetic analysis, however, showed that this view was not just limited—it was wrong, because the microbiome is not only our useful partner, it is vital to our health. The good germs on our skin protect against infection; the good germs in our gut produce vitamins that we must have to live. Children need a broad exposure to the germs around us, which is why Knight co-authored the book *Dirt is Good: The Advantage of Germs for Your Child's Developing Immune System.*

In 2009, Knight had wanted to explore how the skin microbiome changes. Knight, who was at the University of Colorado, planned to swab the mouth, forehead, or arm of one healthy research subject, rub that swab on another person's arm, and then, days or weeks later, see whether the recipient's microbiome had changed. The hazard was not much greater than a mother kissing her child on the forehead, but that is not how the IRB saw it. After its first review, the committee told Knight that it was worried that he might be spreading AIDS.

Twelve years earlier, Princess Diana had arrived at London Middlesex Hospital to open its dedicated HIV/AIDS unit, and during that visit she shook the hand of an obviously sick young man in a wheelchair. She wore a generous smile but no gloves. Doctors already knew that HIV was not spread through social contact, but the photograph of that handshake did much to educate the public and combat HIV ignorance and stigma.

This news had apparently never reached the University of Colorado IRB, which wanted Knight's consent form to warn subjects that his swab could give them AIDS. In response, Knight told the committee that people living with HIV are sick of being treated like lepers. Fortunately, he added, the university's Office of Equal Opportunity had an employee whose job included teaching the clueless about how HIV is—and is not—transmitted, and who would be happy to educate the IRB at one of its regular meetings. The committee declined this offer, and it withdrew its concern about the transmission of HIV/AIDS.

However, on its second cycle of review, it required Knight to inform potential subjects that they could get other diseases, including smallpox. Smallpox is a dreadful illness, certainly, but it is extinct; the last case, worldwide, had occurred almost thirty years earlier. Today the only live smallpox samples are in high-security government labs in the US and Russia. Knight told the IRB that because the research did not include international espionage or time travel, his subjects should be safe from smallpox.

In the end, Knight was allowed to proceed with his research, but the time and effort he wasted dealing with these needless distractions was real, unlike the risks feared by the IRB.

IF IRBS WERE POPULATED PRIMARILY BY ETHICISTS, schooled in the great philosophers but ignorant of science, this fear of HIV and smallpox could perhaps be excused. But the University of Colorado committee, like most IRBs, bristles with practicing doctors and active scientists. It is hard to understand how this group of researchers, some of whom treat and study infectious diseases every day, lose touch with medical reality the moment they walk into the IRB meeting room. Yet they do. And they are not alone, since many other IRBs fall into the same alternate reality, working diligently to protect subjects from risks that are minor or nonexistent.

Let me pause to clarify that while the IRB members and chair are formally responsible for the committee's decisions, we do not actually know that these faculty members were afraid that Knight would spread smallpox and HIV. The concern could have come from IRB staff, whose influence has multiplied since the federal crackdowns of the turn of the century. As a result, decisions that are attributed to the IRB acting in its official capacity (a chair and perhaps ten or fifteen committee members) may in fact have been made by a single staffer. Before the IRB members ever file into a conference room, the staff has answered investigator questions, checked to make sure that proposals pass standard checklists, and passed their own questions and concerns along to the committee. Laura Stark, who sat in on the meetings of three IRBs over more than a year, confirmed that IRB administrators, who are not even members, may still persuade the committee to approve or reject a scientist's proposal.

We should therefore bear in mind that IRB decisions are not always made by the IRB members; in fact, the members may be entirely unaware of many of them. When I refer to IRBs, then, that always includes the staff.

The specific origin of the worry about smallpox and HIV is not, in any case, the point of this story. Concerns like this, in which a remote or impossible risk is used to block or delay a study of the skin microbiome, are a dramatic change from the genuine need and reasonable response reflected in the system's early days. When they were first created, IRBs were designed to protect the public from scientists like Chester Southam, who injected his subjects with live cancer cells without their knowledge or consent. In the modern era, an IRB was alarmed at the (im)possibility that Rob Knight's swab might spread HIV or smallpox from one healthy adult to another. Knight is, unfortunately, in abundant company, for scientists have published hundreds of articles documenting other, equally bizarre rationales for delaying or blocking research.

When IRBs lose track of science, and of reality as most people experience it, they become too willing to block research that poses only modest risks that some subjects might be willing to accept, and too likely to require consent forms that are so bloated that the useful information is lost in the fine print. This aggressive approach slows recruitment of subjects, makes launching research projects more difficult, and directly delays new treatments. The harm done is all but invisible, since the public doesn't know of this lost research. The costs levied by excessive oversight, slowing advances in medical research and treatment, are measured in pain, chronic disease, and preventable deaths.

Fredric Coe, who has been treating kidney stone patients at the University of Chicago for fifty years, is baffled by the way that his IRB has gradually grown to fear risks that are sometimes entirely imaginary. He sums up his experience this way: "Because my research (perhaps out of lack of nerve, ambition, or imagination) involves essentially no risks at all, or risks of so slight a character as to be unimportant, I have been freshly surprised, each time, that IRB requirements are as involved and elaborate as they are."

In one of Coe's studies, volunteers are given a room in the University of Chicago's Clinical Research Center for a three-day visit, where they spend most of their time eating, sleeping, or watching TV. For the experiment, Coe's staff administers a small dose of an intravenous kidney marker (the dose is one-tenth of that used for a kidney X-ray) and takes blood and urine samples. The risks enumerated in the consent form include an allergic reaction to the marker and bruising at the site of the blood draw. (The subjects' time inside the Clinical Center is certainly safer than Coe's on the outside, since his commute takes him, twice daily, into Chicago traffic.)

In order to protect Coe's subjects from abuse, the IRB feels it must review every detail of the protocol, including the timing of the meals and the schedule for blood draws. It approves the experimental schedule, ensures that the consent form is complete, and reviews the signs that Coe uses to recruit volunteers (the signs are authorized only if they bear an official stamp). What's more, the protocol is approved only precisely as submitted. If Coe realizes, mid-experiment, that the science would not suffer if a blood draw scheduled for five a.m. is delayed until after breakfast, he must submit a new packet of forms and wait for re-approval before making the change.

Coe is resigned to IRB review, but he sees a bizarre imbalance between the effort required to get the protocol and the consent form approved by the IRB and the modest risk his subjects run. He sees "little or no risk, accompanied by massive IRB requirements serving an unclear purpose. ... Where did the idea come from that any risk at all, any element howsoever slight, minute, and atomic in scale, needs oversight?"

This level of supervision would be more appropriate for an inmate on work release from prison than for a doctor who is licensed to practice medicine by the State of Illinois. But regulators, including

IRBs and particularly their federal overseers, believe that their stringent requirements are essential to protect subjects. To admit that the risks they so zealously guard against might be minimal would undermine their legitimacy, and so they argue that minor risks threaten major harms.

Those risks now include the disclosure of genetic information, which is becoming increasingly accessible. People who were conceived through sperm donation can use 23andMe, Ancestry, or MyHeritage to identify their biological fathers, and detectives can use DNA extracted from a cigarette butt or coffee cup to identify a criminal suspect. Scientists create DNA databases that could be hacked and, at least in theory, lead to the identification of individual subjects along with information about their physical characteristics and vulnerabilities to disease. In response, Congress has made it illegal to discriminate against job applicants, employees, or applicants for health insurance on the basis of genetic information. Other protections are evolving, including those that address the privacy of research subjects in genetic studies.

Harvard psychologist Steven Pinker is among those who have pushed back against this attempt to prevent largely-theoretical harms. Pinker feels that focusing on hypothetical problems misses the promise of "vast increases in life, health, and flourishing" of new technologies like CRISPR gene editing. His conclusion: "Given this potential bonanza, the primary moral goal for today's bioethics can be summarized in a single sentence. Get out of the way."

These issues do not concern Fredric Coe. His research requires no genetic information, his lab doesn't have the equipment to analyze the DNA in the samples he collects, and he holds those samples in secure conditions. To discreetly find the DNA sequences of a Coe patient who is suspected of a crime, a detective would need to loiter by the door of the kidney stone clinic. Perhaps the patient will, after leaving the clinic, stop by the Plein Air Café and leave behind a cigarette butt or coffee cup.

Leaving genetics aside, a morbid imagination makes it easier to imagine otherwise obscure dangers. In her influential 1992 manual, *Planning Ethically Responsible Research*, Joan Sieber teaches IRB members how unexpected risks may materialize, using as her example an exercise physiology study. Such investigations include, for instance, studies of the uptake of oxygen and expenditure of calories by a subject running on a treadmill. In such an experiment, Sieber notes, an injury might be physical, such as a black eye. It might be psychological, with the subject suffering a black eye "becoming depressed about attending events wherein wrong inferences may be drawn about the cause of the black eye." Or social: "being rejected for having that black eye." Or economic: "being passed over for employment in favor of another candidate who interviewed without a black eye." Or even legal: "being arrested and interrogated about the possible connection between the black eye and a brutal assault that left a neighbor comatose." Sieber concludes that "these examples illustrate risks to subjects," even though they are drawn not from any actual events but from her imagination.

I think the consequences are unlikely to be so dire. Using my own imagination, I see the research subject, who has perhaps been recruited because he is a student-athlete, exhausted from his sprint on the treadmill but happy to have made his contribution to sports physiology. The scientist is also pleased—she has the data she needs and will soon finish her thesis. She watches as the treadmill slows and stops. Then the young man, stepping off the treadmill, stumbles and strikes his eyebrow against one of the machine's supports.

"Ouch!"

"Are you all right?" The scientist stretches out a hand to steady him.

"I'm going to get a shiner for sure. But don't worry—I've been hurt a lot worse playing rugby."

"I'm so sorry! Let me get you an ice pack."

Half an hour later, the subject learns that the scientist, who is apparently single, hopes to treat him to lunch. "It's the least I can do," she says, and our picture fades.

To adapt Sieber's phrase, couldn't this happy outcome also illustrate benefits to subjects? Not for IRBs, who do not allow their imaginations to consider the positive side of the equation. Instead, whether they are located in the United States or other English-speaking regions around the world, they follow Sieber's catastrophizing ruminations. Maureen Fitzgerald, who observed IRBs in five countries, reports that they construct "the what if or worst case scenario narrative. ... This kind of narrative is generative, as each potential version of this story builds on the one before it, and each version becomes more and more serious until it gets to the worst scenario they can come up with, one that may significantly overestimate the kind, potential for, probability of, or seriousness of the risk." These scenes can "take on lives of their own. They can develop to the point where an uninformed listener might wonder if this research was worth the risk because as the versions of the hypothetical develop they become more and more believable and members with a moral conscience are placed in a position where they feel they have to raise questions about whether or not the project should be approved ..."

Psychologist Jonathan Baron points to another factor favoring this approach: "Members of review panels must demonstrate to the other panel members that they are diligent in doing their jobs, which is ostensibly to protect subjects from harm. The easiest way to demonstrate diligence is to read the proposals that come in and look for possible risks that could be avoided or about which subjects could be warned."

IRBs' reluctance to allow potential subjects to accept risk is almost inevitable. They are charged with predicting whether unmeasured risks and uncertain benefits are appropriate for subjects whose preferences are equally unknown. It's an almost impossible task that involves weighing multiple imponderables. This mandate also ignores the fact that, in many settings—not just in research—people often willingly accept risks that they do not wish to be responsible for another person taking.

The tendency toward ultra-conservatism is magnified by group dynamics. Committees that make decisions about the welfare of third parties are likely to be particularly risk-averse. IRBs thus tend, in ethicist Michelle Meyer's summary, "through a broad understanding of research risk and a narrow understanding of research benefit, to make risk-benefit decisions that reflect the (imagined) preferences of the most vulnerable, risk-averse participants."

Consider, as a concrete example, three people who are considering enrolling in an experiment that requires them to have a lumbar puncture (a spinal tap). It's a procedure each candidate has had at least once in the past, so they have personal knowledge of the procedure. One of these people, on learning that a lumbar puncture is involved, will say "No" with a shudder of horror. Another, who might have had a lumbar puncture with excellent local anesthetic, might join if the study offers sufficient payment. The third might learn that the study is struggling to recruit enough participants, see the work as important, and decide to join on this altruistic basis.

Three individuals, three different decisions—each equally valid, in my view. The decision as to whether lumbar puncture is too burdensome should be made by the potential subjects, not the IRB.

THE ULTRA-CONSERVATIVE APPROACH TO RISK also impacts the way consent forms are handled. Scientists have some input into the content of consent forms, but IRBs have the final say about what the forms must say, how they must say it, and what they must omit. Because IRBs are deeply afraid of running afoul of OHRP, they generally insist that consent forms minimize benefits and exaggerate risks. This gives potential subjects an unbalanced picture of the research.

To be sure, IRBs' caution about scientists overselling research has a valid historical basis. Recall that Chester Southam led his subjects to believe—falsely—that his cancer cell injections tested their immunity in a way that might help in their care. In 1970, Paul Ramsey, professor of religion at Princeton, wrote insightfully about scientists' unbalanced view of their own work. In his classic *The Patient as Person*, Ramsey did not mention any investigator by name, but he captured with precision the loss of objectivity exemplified by Southam's behavior. Ramsey also proposed a solution: The scientist can approach balance by "leaning against" his or her bias in favor of the research. The way to do this is by exaggerating his or her sense of the risks and discounting the benefits. This will shift the scientist away from his or her biased perceptions and toward a more balanced or objective position. Ramsey's advice was useful—but it was intended for the scientist, not the IRB.

The exaggeration of risks and the minimizing of benefits create a barrier to informed decision-making for potential subjects. However, they serve the IRB's purposes well. An ultra-conservative presentation of risks and benefits makes it less likely that a disappointed subject will complain. It also makes it less likely that OHRP will punish the IRB. If something goes wrong—if a subject is seriously injured or dies—federal authorities will examine the consent form for any trace of optimism, which will be seen as an indication that subjects were duped into participation

In contrast, OHRP has never criticized an IRB for approving a consent form that presents this distorted picture. Thus, all the incentives push IRBs and the researchers who ask them for permission to tilt consent forms toward emphasizing the negatives and minimizing the positives. This biased approach deprives potential subjects of balanced and accurate information about risks and benefits, making it more difficult for them to reach a decision that matches their values and preferences.

Let's consider a common and important example. Randomized controlled trials, like Melinda Moir's study of codeine versus placebo, are the workhorses of modern medical science, and they provide a useful lens through which to understand the risks and benefits of enrolling in research.

Because it is unknown which subjects will receive a particular treatment (or any treatment at all), any given patient may do better or worse than they would in routine clinical care. We cannot know which is more likely ahead of time—that is why the studies are done. Yet we do know something about whether or not an average patient would be smart to enroll in a typical trial.

This issue has been debated for years. Philosopher Hans Jonas was deeply skeptical of double-blind trials, particularly those that include a placebo, which he considered "an outright betrayal of trust in regard to the patient who believes that he is receiving treatment. ...The patient is definitely wronged even when not harmed." In 1974, Harvard law professor Charles Fried, noting that treatment in a study follows a standardized algorithm, argued that even subjects who know they are in a randomized trial would be better off being treated by their own doctor.

More recent scholars have shared this opinion. In 2006, Jerry Menikoff, a doctor and ethicist then at the University of Kansas, wrote a book with Edward Richards providing examples of people with serious diseases—often cancer—who enrolled in a trial and later regretted that decision. Menikoff saw these unfortunate cases as examples of a universal problem in research, which "involves intentionally exposing persons to risks, and not for the primary purpose of treating them or making them better, but rather to answer a research question. ... We are back to a classic dilemma: society is regularly involved in determining how 'suffering shall come to some persons and not to others.' These are the tragic choices involved in designing a system for research on human subjects."

Of course, not all experimental subjects suffer; Menikoff conceded that "thousands of people have had their lives saved precisely because they enrolled in research studies." But, he went on to say, this does not mean that study participation is on average a good idea, since every trial involves the imposition of extra risks. These extra risks include not only the possibility that the treatment received in the trial will be inferior, but also risks that "may be very small. They might not lead to any harm, or might merely involve a small amount of inconvenience to the subject."

I quote Jerry Menikoff not because I agree that "a small amount of inconvenience" helps illustrate society's "tragic choices," but because, since 2008, he has been the head of OHRP.

One weakness of Menikoff's argument is that he sees inconvenience as evidence of tragedy. On a deeper level, he supports his position with anecdotes and disregards the abundant evidence about how patients actually fare when they decide to enroll in research. Jeffrey Peppercorn, for example, reviewed cancer studies, comparing the outcomes of patients treated in or out of clinical trials. He found seventeen relevant studies. Five of these showed no difference, and in the remaining twelve some or all of the trial participants benefited. He chose a "cautious interpretation," and did *not* conclude that trial participation was better than routine care. This brings to mind the way that consent forms for controlled trials perseverate on the risks, and minimize the possible benefits, of research participation. Peppercorn's analysis suggests that, for cancer patients, this is the wrong approach. Gunn Vist, who did a thorough review of relevant data in 2008, likewise concluded that there is no proof of either harm or benefit from trial participation.

Some ethicists believe that this conclusion is too cautious. In their view, subjects in research often benefit from a "trial effect." This means that they may be better off, in either arm of a trial, if they are being cared for by doctors with a special interest in their disease, following a carefully-thought-out protocol, rather than by a doctor who is providing routine care. Pediatrician John Lantos therefore asks, "What if, instead of creating increased risk, clinical research creates increased benefit? Perhaps we should include, as part of the informed consent process for clinical research, a statement to the effect that participation in a research protocol has been shown to lead to better outcomes than nonparticipation."

Pediatrician Norman Fost agrees. He does not "think the public yet appreciates how desirable it is to be part of a research study—that is, particularly if you're sick. Your chances of being well taken care of, and reducing your risk of harm, and achieving benefits, I think, for a lot of conceptual reasons and empirically, are much higher if you're in a well-designed, well-supervised research study than being in a doctor's office."

THE DEATH PENALTY ERA ENDED IN 2001, but OHRP continues to demand that IRBs take an ultra-conservative approach to risk. The agency itself has shown a striking ability to ignore the benefits of research participation and exaggerate the risks. The most extreme example came when it temporarily halted a study, conducted by Johns Hopkins Hospital intensivists Sean Berenholtz and Peter Pronovost, that was showing how to save lives in intensive care units.

The ICU is where hope battles against onrushing disease. Patients suffering from sepsis—a severe reaction to a severe infection—are best treated in an ICU, where their mortality will still be between 20 percent and 40 percent. Patients with severe Covid also require ICU care; their mortality, early in the pandemic, was 40 percent. When a hospitalized patient takes a turn for the worse—when death is suddenly a likely outcome—the ICU is their last hope.

Whatever condition brings them to the ICU, patients there are constantly vulnerable to complications that result not from their disease but from their being in the hospital. Patients die, for instance, when a central line—a large intravenous catheter whose tip floats near the heart—becomes infected and bacteria pour into the bloodstream. In the mid-2000s, the number of central line infections in the United States was about 80,000 per year, with deaths estimated to be as many as 28,000. These infections are often caused by a doctor who inserts the line without following every step in this routine checklist:

1. Wash hands
2. Sterilize the skin
3. Cover the patient with a sterile drape
4. Wear a hat, mask, and gown
5. Use sterile gloves
6. Apply a sterile dressing

In 2004, this list had not changed for decades—nor had the rate of infection and death. Berenholtz and Pronovost began addressing the problem by examining whether doctors were actually following the list consistently. In the first phase of their project, they quietly watched, and they discovered that doctors often were too rushed to follow the list, or found the necessary supplies were out of stock, or just forgot one or more steps.

Next, Berenholtz and Pronovost created a special cart that contained all the necessary equipment and supplies, and provided ICU nurses with a checklist of the six infection-control steps. The nurses were told to monitor each central line insertion and to stop any doctor who was about to skip a step. This was a change in dynamics, because the nurse was now expected to say, "Excuse me, doctor, but I haven't seen you wash your hands. Would you mind doing that before going any further?" This experiment was not testing a new medical procedure; rather, it attempted to get harried doctors to slow down enough to do what they should have been doing already.

In the experiment, the percentage of cases in which the doctor followed every single step of the checklist rose from 62 percent to 100 percent. Of vastly more importance to patients, when the new protocol was made routine at John Hopkins, the infection rate fell to zero. Based on the usual frequency of fatal infections and the number of intensive care patients, Berenholtz calculated that each year this intervention prevented an estimated forty-three infections and eight deaths, and saved almost $2 million in additional medical costs.

Could these results be achieved at institutions that were not under Pronovost and Berenholtz's watchful eye? And at different kinds of institutions—community hospitals, for instance, as well as teaching hospitals? Pronovost persuaded the Michigan Hospital Association to collaborate with him in a larger trial that included smaller hospitals that do not ordinarily participate in research, as well as struggling urban hospitals in which funding, supplies, and manpower are all in short supply (think Detroit).

The Johns Hopkins IRB reviewed the protocol. It determined that the study was exempt from IRB oversight because the data provided to Pronovost's team did not identify individual patients. In 2007, after the study was well underway and lives were being saved across the state, OHRP decided to investigate. It concluded that the study was not exempt, that consent should have been sought from the subjects of the research, and that in failing to obtain their consent Johns Hopkins had violated federal regulations. And who were the research subjects whose consent OHRP wanted obtained? The patients in the ICUs. And the nurses who were keeping track of the checklist. And the doctors who were inserting the catheters. Even though the point of the experiment was to get the doctors to do what they were already supposed to be doing.

OHRP demanded a corrective action plan. It appeared that the study could not continue until consent was obtained from every patient, nurse, and doctor, with each hospital's IRB supervising that process. OHRP showed no concern about the small hospitals that had no IRB, let alone the cost to patients who would be deprived of better care.

Atul Gawande, a respected doctor, best-selling book author, and essayist, led the defense. In an op-ed in the *New York Times*, he wrote that OHRP was "in danger of putting ethics bureaucracy in the way of actual ethical medical care. The agency should allow this research to continue unencumbered. If it won't, then Congress will have to."

OHRP was caught off guard by the criticism and perhaps by the idea that its aggressive policing might not win universal praise. When Ivor Pritchard, the agency's head, was pressed for an explanation, he argued that there was a risk that the checklist would slow down care, which he presumably considered a bad thing. In addition, he said, obtaining consent would protect subjects from harm because "there is a risk that the quality of care could go down." How could the quality of care be reduced when fewer people were dying? Pritchard had a theory: Having nurses challenge doctors who were not following the checklist could create animosity. "That's not likely," he said, "but it's possible."

Ethicist Ruth Faden commented that this study, which was "saving thousands of lives by preventing central line-associated bloodstream infections ... was almost halted due to concerns about research ethics oversight. But few have come forward to express concerns and oversight for the thirty thousand or so people who will die unnecessarily each year in the United States from this type of infection."

Managing the risk involved in medical experimentation is an important consideration when designing any research protocol. But this challenge demands a thoughtful approach that recognizes and balances both the risks and the potential benefits—an approach that is impossible today.

# 5

# Complicated Consent

L ET'S SAY THAT YOU ARE NAVIGATING an icy sidewalk, slip, and fall on your outstretched hand. Your hand is all right, but your clavicle (collarbone) hurts, a lot. You go to the emergency room, where X-rays confirm that it's broken. The doctor puts you in a sling and tells you to follow up with an orthopedic surgeon.

Orthopedists are not the dithering sort: They tend to believe that for any given type and location of fracture, there is one best treatment. This conviction may obscure the fact that orthopedists sometimes disagree on what the "one best treatment" is.

For your particular fracture, Orthopedist A may explain that while surgical repair is a theoretical possibility, you will do better if you simply wear an appropriate sling and allow the body's healing power to work its wonders. If you ask whether an operation might be better, you may be told that while it is possible to surgically connect the broken ends, surgery has its own complications and patients treated in a sling have a high rate of healing.

So far, so good. But if you happen to visit Orthopedist B, you would learn that a clavicle that heals on its own may be significantly shorter, which causes weakness on that side and pain when you reach overhead. Further, the "high rate of healing" without surgery is about 90 percent, so one patient in ten ends up needing an operation anyway.

For our purposes, we will assume that you visit Orthopedist C. She explains that since her colleagues do not agree on the best treatment, she is participating in a randomized trial of the two approaches, surgical and nonsurgical, and you are eligible. If you have a preference of your own—if, for instance, you are terrified of surgery, or if your cousin had the same fracture, had an operation, and did very well—then she will do as you prefer. If you have no earthly reason for preferring one treatment over the other, then you are eligible for the trial. If you agree to enroll, a random process, like flipping a coin, will determine which treatment you will receive. You say you would like to join the trial, and she reaches into a drawer and pulls out a consent form.

Fredric Coe has walked patients through a similar process many times. When talking with a patient who might be willing to participate in a kidney stone study, he retrieves the IRB-approved consent form, which describes the study's purpose and duration, the procedures involved, a list of risks (ranging from possible and serious to unlikely and trivial), a caution that participation will not benefit the subject, an explanation of whom to contact in case of questions, a reassurance that its subjects' identities will not be disclosed in any publication and that the University of Chicago will keep all data private except as required by state and federal law, and much more. The IRB asserts that every word is essential, but few potential subjects agree. The patient sees at a glance the nature and approximate contents of this document and begs to be spared. "Doc, let's get by all the words ... I trust you, not that stuff on the paper." As sociologist Carol Heimer writes, "Apparently research subjects treat informed consent as yet another bureaucratic routine, not so much part of decision-making as part of implementing a decision already made . . . ."

Back to Orthopedist C and your fractured clavicle. The IRB believes that the consent form contains information that is vital to your decision about enrollment. But I believe that your choice is more likely to be determined by your preferences and predispositions, since you decide whether or not to enroll in a clinical trial just as you make other decisions.

You bring to the choice all that you know and have experienced, not just what you discuss with the orthopedist or read on the form. You bring your preexisting attitudes and opinions, your knowledge about medicine and science—whether based in fact or myth—your beliefs about religion and hospitals, and your convictions about how society functions and how it can be improved. You may trust this orthopedist, this hospital, and the government to act in good faith, or you may not. Your attitudes toward medical research may pair with your attitudes toward clinical medicine: You may believe, for instance, that vaccinating children saves lives or that vaccines are a threat to children's health and parents' rights.

I believe that you have a pretty good idea whether you are going to participate in the fracture study before you learn anything about its details, and that your decision may have scant connection to a printed consent form whose primary purpose is to serve the institution's interests.

In this book, I discuss "consent," not "informed consent." The latter phrase is useful in a legal context, since American courts usually consider patients to have given their informed consent to clinical treatment when they sign a form that includes a specific set of information. But for our inquiry, the word "informed" adds little, for every voluntary experimental subject is informed to a greater or lesser degree. For our purposes, the question is not whether consent is "informed" in the legal sense but whether, in the real world, prospective subjects know enough to make a decision that is right for them.

The meaning of "consent" is itself in dispute. Ethicists generally believe that consent begins with a conversation in which the scientist provides information and potential subjects ask questions to clarify any uncertainty. Prospective subjects then go home with written information that they can mull over and review with their family, their doctor, or their lawyer. Thus, they can make an unhurried and thoughtful decision that fully reflects their values and preferences, a decision that they are free to later change. "Consent" is this entire process. When there is a written form, that form is not the consent, it is only a document confirming that the process took place.

In practice, however, this theoretical definition is ignored by the people who actually oversee and conduct consent, since both IRBs and scientists disregard the process and focus on the form. In doing so, IRBs are eminently practical, for they could not possibly prescribe and oversee the conversations that thousands of scientists have with an even larger number of potential subjects. Nor could they ensure, as ethicists would like, that those countless potential subjects engage in thoughtful conversations with trusted advisors and then make careful decisions about participating in research.

In contrast to this conversation, a written document is a straightforward tool that is subject to tight control, with every word reviewed and approved. Consent is thus telescoped from a conversation to a form, and while its nominal purpose is to protect research subjects, its major function is actually to protect against two threats: audit by OHRP and litigation by an unhappy subject.

Unfortunately, a consent form that is optimized for protection of the institution takes longer to read than one designed with the subject in mind, is more difficult to understand, and slows research enrollment. Both potential subjects and the research itself suffer.

The first and most obvious problem with these protective consent forms is length: Twenty or thirty pages is common for consent forms today, and some forms are even longer. Forms that are so long they are difficult to read and understand pose a problem for potential subjects, but they are required by the system.

Shannon's 1966 policy instructed IRBs to ensure that consent was "appropriate," but no written form was required. By 1974, it was assumed that a form would generally be used, and regulations issued that year required the form to include six specific elements: the procedures to be followed, any expected discomforts and risks, any likely benefits, any alternatives, an offer to answer questions, and formal reassurance that the subject could withdraw without prejudice at any time. Revisions in 1981 added statements about confidentiality, compensation for injury, contact information, and six optional additional elements.

By 1985, American consent forms resembled the Michelin man. Suppose, for instance, that you went to the emergency room in 1985 with a heart attack, and you were offered enrollment in the Second International Study of Infarct Survival (ISIS-2). The elements required since 1981 would all be present, and now the form would also include a notification that "your medical record ... might be inspected and/or photocopied by the Food and Drug Administration or other Federal or state government agencies in the ordinary course of carrying out their governmental functions" and another that "[i]f your record is used or disseminated for government purposes, it will be done under conditions that will protect your privacy to the fullest extent possible consistent with laws relating to public disclosure of information and the law enforcement responsibilities of the agency."

These statements are true. Whether they help a potential subject decide about enrolling is another matter.

Other elements have since been added, and the length of each has increased as well. These longer forms become more difficult to understand. For example, the consent form for the SUPPORT study of the optimal level of oxygen for extremely premature babies, written in 2005, begins with an explanation of the science behind the research. The first paragraph explains that "a baby's lungs are made up of tiny air sacs" exchanging oxygen and carbon dioxide, and that in premature infants those lung sacs "don't always work this way." It goes on to review the options that doctors have to treat immature infant lungs, including medication. Another option is to "place a resuscitation bag over the baby's nose and mouth to provide oxygen and manual breaths," which could use "continuous positive airway pressure or CPAP or PEEP."

There is much more, a total of nine single-spaced pages. Each sentence, read in isolation, may be clear, but page after page of complicated medical information is likely to baffle even an eager learner. Legal scholars Omri Ben-Shahar and Carl E. Schneider put it well: "Complexity cannot be explained simply. Sophisticated vocabularies and professional languages encapsulate complex thoughts. If only simple words can be used, everything must be lengthily spelled out. . . . Many words, even if simple, make forms repellently long and cognitively overwhelming."

In the 2010s, OHRP revised its regulations in response to perceived difficulties with the IRB system. The regulations that resulted, which were finalized in 2018, note the problem of lengthy forms. They also require the addition of *more* explanatory material, which seems unlikely to solve the problem.

CONSENT FORMS WERE ALREADY TOO LONG back in the 1960s. Johns Hopkins professor Louis Lasagna hoped that evidence might bring reason to regulation, so, at the same time he was jousting with Henry Beecher and Hans Jonas at the AAAS conferences, he conducted an experiment that probed whether it is possible to give experimental subjects too much information.

Lasagna and a colleague, Lynn Epstein, invited people who worked in their office building to read a consent form about a trial of a new medication for headache and to answer a few questions about it. There were three versions of the consent form. The short form was three snappy paragraphs, the intermediate form was more than twice as long, and the long form was longer than the other two combined.

The more information subjects received, the worse they did on the post-consent quiz. Subjects who read the short form did best, but even they answered, on average, only 67 percent of the questions correctly. Subjects who read the intermediate-length form got 45 percent correct. Those who read the long form, with its comprehensive list of side effects, scored only 35 percent. Of special concern were the subjects (using the medium and long forms only) who failed to recognize that their medical histories made the drug particularly hazardous for them. None of the short-form subjects made this mistake.

As Lasagna summarized the results, "lengthy, detailed expositions of risks and purposes may defeat the process of communication, with less comprehension of the problems and dangers than if one uses a brief, straightforward statement." He noted that this spotlighted a problem with the 1966 FDA requirements, which "can be interpreted to mean that 'informed consent' will be most validly obtained after a lengthy and detailed discussion of the actions and toxicity of the drug under consideration." However, if "valid" meant that subjects reached maximal understanding of the research and could use the information to keep out of harm's way, then shorter was better.

Lasagna and Epstein's conclusion that the harder you try, the less consent educates, has proven durable. The literature between then and now abounds in studies that attempt to bring potential subjects to deeper understanding by using either an improved written form or additional educational material, such as videos and anatomy models. Traditional methods for enhancing consent (better formatting, more white space, bulleted lists) and newer approaches (such as interactive instructional videos) do make some difference, but it is small, almost negligible. I suspect this is partly because, despite the effort lavished on these improved presentations, they are so boring.

The failure to inform—or, rather, the failure of potential subjects to understand—can reach impressive levels. Consent forms usually begin with an invitation to the potential subject "to participate in a research study" or an equivalent phrase. They explain what the experiment will involve, reassure potential subjects that they do not have to participate, tell them that they can withdraw at any time, and then go on to a lengthy recitation of facts and figures. But even when a trained educator goes over the consent form line by line with the potential subject, and they discuss any remaining questions or issues, and the patient nods, says "let's go ahead" and signs the form, you can ask them a week later if they are in an experimental study and some will tell you they are not. The comprehensive consent that is today the norm performs other tasks, such as educating subjects about the nature of the research, its risks, and its benefits, even more poorly.

Today's federal regulators are well aware that institutions see the primary goal of consent forms as protecting the institution. In an attempt to remedy this awkward reality, the regulations seek to prohibit institutional self-interest in consent forms, forbidding "exculpatory language through which the subject . . . is made to waive or appear to waive any of the subject's legal rights, or releases or appears to release the investigator, the sponsor, the institution or its agents from liability for negligence." This fig leaf is too small, since the long list of study risks is transparently intended to protect the institution from federal officials and angry subjects. A signed consent form warning of the very harm that materialized may be the IRB's best protection from the critical gaze of an OHRP auditor or trial judge.

This reality rankles ethicists like George Annas. Consent forms, he writes, were meant to serve as a shield to protect subjects, but regrettably the lawyers who draft them "may view . . . the research subject as an adversary" and "the institution or researcher as their client." Exactly! It is the institution, not the research subject, that pays the lawyer's salary. Annas, who is a professor of law, surely knows that a lawyer who does not protect their employer will soon be out of a job.

Prospective subjects are not idiots, and most immediately recognize the consent form's true purpose. Oncologist Christopher Daugherty, who has served as chair of the University of Chicago's IRB, interviewed 144 subjects who were enrolled in cancer trials. He found that a majority believed that the consent form was intended to protect the scientist and the institution from liability.

But we should not blame the lawyer, the IRB, or the institution, for they are responding to OHRP's unbalanced oversight. Even OHRP itself deserves only a share of the blame, for the agency operates within the constraints established by the National Research Act of 1974 and the system of unbalanced review that it created.

Let's return to Coe's office and his conversation with his patient. Keeping in mind today's bloated consent forms, it's unsurprising that Coe's patients want to skip what they see as worthless paperwork. They intuit that its primary purpose is to protect the institution from liability, not to protect them from harm or help them make an autonomous choice based on their values.

However, the paperwork is more than an extraneous annoyance; it imposes real costs, particularly in the form of time. For Coe, that time begins when he draws up a consent form to submit to his IRB. It takes time for committee members to review the form, time to discuss it at the meeting, time to return it for modifications, time for Coe's patients to read and sign it, and time for his staff to file it and produce it on request in case of an audit. This investment of time would be right and proper for research that imposes significant burdens or carries appreciable risks, but it is a waste for the low-risk studies that are Coe's stock in trade. Just the kind of studies that compose a significant fraction of all medical research.

LET'S BROADEN OUR INQUIRY beyond consent forms and draw in other aspects of IRB review. This will take us to one of the most disturbing aspects of today's IRB dysfunction.

# 6

# Lost Lives

I RB REVIEW COSTS LIVES.

This is not necessarily a bad thing. The unregulated boom in medical research after World War II achieved remarkable scientific results, but at the cost of research subjects who were enrolled in research with too much risk and too little consent. Here I am thinking of Nathan's shortened form of consent ("You are the patient. I am Doctor Nathan. Lie down.")

When James Shannon introduced the IRB system, he knew that it would block some promising but abusive studies, and he accepted that. He optimistically predicted that the IRB system would serve to keep the government (and its scientists) "out of trouble," while at the same time it would also "encourage the flourishing of sound clinical investigation rather than discouraging it." Unfortunately, the discouragement he feared is here, and its costs are out of control. The solution is not to abolish regulation, for oversight is and will remain essential, but to reform it to bring its costs within the bounds of reason.

Some costs must still be accepted, if only because the ordinary process of oversight is bound to cause some delay. That means that the development of new treatments will be delayed, and some patients will suffer or die as a result. If those running the oversight system remain ever-conscious that delay imposes hardships on patients who need better care now—so long as IRB review is efficient—then the number of lives lost will also be minimized.

This consciousness is absent from the IRB system's current operations, with calamitous results. As British ethicist Julian Savulescu writes, "There is a moral imperative to perform good research and not unnecessarily impede it. To delay by 1 year the development of a treatment that cures a lethal disease that kills 100,000 people per year is to be responsible for the deaths of those 100,000 people, even if you never see them."

Defective IRB review can cost lives in a variety of ways. Let's consider three examples, one each from Australia, the United Kingdom, and the United States. The three countries have similar oversight systems, although in Australia and the United Kingdom the reviewing panels are usually called ethics committees and the process is called ethics review.

In the early 2000s, Australian oncologist D.R.H. Christie became increasingly frustrated as his potentially life-saving cancer treatment studies crawled through ethics review. With colleagues G.S. Gabriel and K. Dear, he surveyed other cancer centers. All reported significant delays in the review process for research, with an average of two months' wait.

Overall cancer survival is improving by one to two percent per year, thanks largely to the discoveries made through research. Ethics review slowed that research and the health benefits it produces. Thus, in Christie's opinion, ethics review should be charged in the deaths of the people who die before better treatments reach their bedside. Christie calculated that the two months' delay in research approval was responsible for about sixty unnecessary deaths in Australia each year. The details of Christie's analysis can be debated, but there is no doubt that slow oversight costs lives.

Even prompt IRB review can cost lives, as when it interferes with a study's methods. In the early 2000s, British sociologist Robert Dingwall was asked by the National Health Service to find out why doctors and nurses were re-using needles and IV tubing that were supposed to be used once and then discarded. Re-use is strictly banned, but this prohibition was sometimes ignored, leading to the spread of HIV/AIDS and hepatitis. The NHS knew that this practice was responsible for many infections and an estimated seven deaths every year, and asked Dingwall and a colleague to find the reason.

It can be difficult for health care providers to admit that they are breaking the rules and putting patients' lives in jeopardy, but social scientists have long experience exploring deviant behavior using properly-constructed surveys. Part of Dingwall's plan was to send such a survey to "the relevant staff in about 350 hospitals." As I read the article, I remembered going to the post office to mail a couple of boxes of my own surveys and thought that Dingwall would be doing the same. It was unfortunately not so simple. Before the surveys could be mailed, each of the hospitals needed to approve the plan, which Dingwall estimated would generate "about 1600 signatures and 9000 pages of documentation." Dingwall's colleague would also have to undergo "around 300 occupational health examinations and criminal record checks."

Dingwall supports research oversight for biomedical experimentation. In view of its "very serious risks ... it is entirely proper that investigators should not be judge and jury in their own cause." But the poor fit of British oversight with his research forced him to pare back his study, which reduced his ability to counsel the NHS on how to stop the spread of infections. In frustration, he wrote that "the ethical cost of the NHS system can be measured by the lives that will not be saved."

Christie objected to the lives lost due to the time consumed in the ethics review process; Dingwall objected to the lives lost as a result of the interference in his methods. In our third example, lives were lost because of the delay caused by a cumbersome consent process. It was a study of new treatments for heart attacks. The study was called ISIS-2 ("Second International Study of Infarct Survival"), and Rory Collins and coauthors Richard Doll and Richard Peto were among the lead investigators.

I want to locate this study in the larger context of the struggle against heart disease and stroke. I was born in 1947, and when I was growing up it was commonplace for people in their fifties to drop dead of a heart attack. Now it is a surprise when heart disease kills someone in their sixties. The data bear this out, for the progress we've made against this scourge is almost incredible, with the death rate from heart attack and stroke falling by two-thirds between 1950 and 2000.

This plunge in the death rate had multiple causes. Cigarettes kill more people through heart attacks than lung cancer, and the campaign against smoking reduced both diseases. Better drugs for high blood pressure and cholesterol improved these risk factors, as did better lifestyles. More people are exercising. Most dramatically, a series of major studies revealed that it is possible to intervene in the middle of the heart attack itself.

ISIS-2 was one of those studies. In the first minutes of a heart attack—as soon as the patient showed up in an emergency room—the study's experimental treatment sought to remedy the immediate cause, the clot or thrombus in one of the arteries that supplies the heart with blood. As late as the 1970s, cardiologists prescribed oxygen, morphine, and bed rest for heart attack patients, but they had no treatment for the clot itself. Then early experiments suggested the possible benefit of thrombolytics—"clot busters." These drugs included aspirin, which was in routine use for everyday aches and pains, and streptokinase, which had been developed as a treatment for inflammatory lung conditions.

ISIS-2 was designed to test the power of such drugs to save heart attack patients. Subjects who enrolled in the study would be chosen at random to receive one or both of the treatment drugs or a placebo, so that one-quarter received each of the four possible combinations:

aspirin and streptokinase, aspirin and placebo, placebo and streptokinase, or two placebos.

The trial was ethically sound because experts disagreed about the value of the thrombolytics. At the time it launched, in 1985, some cardiologists had begun to use these drugs, but not everyone was persuaded of their value. While aspirin was considered relatively harmless, some experts condemned streptokinase as dangerous and ineffective. As a result, some cardiologists prescribed one or both drugs, but most did not, resulting in a kind of lottery in which the treatment patients received depended on whether the doctor they happened to see used thrombolytics. Patients who were admitted to the care of a cardiologist who felt thrombolytics were dangerous were not informed of the controversy, nor were they asked for their consent for the medications to be withheld. Conversely, if their cardiologist believed that thrombolytics were saving lives, they were not asked to consent to their administration. In either case, consent played no role.

I saw this dynamic during my residency in the early 1980s, when I worked in a hospital with about a dozen cardiologists. One or two of them gave blood thinners like warfarin to every heart attack patient; the rest never used them. The cardiologists who used the blood thinners never asked for consent; the ones who didn't, acted as if this were the only reasonable treatment and did not ask for consent either. And although there is an argument to be made that the patients should have been asked for consent, bear in mind that they had been rushed to an emergency room where the staff was doing all it could to ease their pain and fear and increase their chance of survival. A fresh heart attack is no time for a leisurely consent process, especially if the doctor thinks there is only one best choice.

This, then, was the normal practice when ISIS-2 was launched. The study was conducted in sixteen countries, and each country chose its own consent process. At the time, the United Kingdom, which took the lead in the trial, had national health care but no national ethics oversight. Cardiologists providing routine care did not request permission to give or withhold thrombolytics, and they offered little more choice to people who were eligible for enrollment in the trial.

Unless a local oversight committee required more, UK study leaders required doctors to provide only, in Collins's words, "some vague mention of the trial intended merely to offer patients an easy opportunity to *initiate* any discussion they want." They were prepared to give additional information when subjects or their families wanted to know more, and to exclude a patient who objected to the idea of research. Patients who did not object, or were not paying attention, were in the trial. This disclosure method was acceptable consent by pragmatic British standards of the time.

The idea that a patient could be enrolled in a study of heart attack treatments through this wisp of consent—a mere mention of the research—would shock a contemporary American ethical consciousness. Yet I could discover no complaints by British subjects who had been enrolled in ISIS-2 without reading and signing a form.

American regulators took the opposite approach. The consent form approved by the Harvard IRB, for instance, presented detailed information about every aspect of the study, including the drugs involved. In its discussion of aspirin, the consent form specified that the drug has risks (including stomach upset and an increased bleeding tendency) and noted that these risks might be less in the trial because the dose to be given was only half a tablet and it was coated to reduce stomach irritation. Potential subjects, all receiving emergency care in the early stages of a heart attack, were told that they would be asked to chew the first aspirin tablet and warned that they might find it "distasteful." The form also included all the other elements of consent required by the regulations plus Harvard's institutional boilerplate. Setting the irrelevant material aside, it included an impressively comprehensive description of the experiment, one that few laypeople would be able to read and understand—and vastly fewer who were in the midst of a heart attack.

Patients who made it to the end of the four pages of single-spaced information had one more paragraph to read before signing:

"I have been fully informed as to the procedures to be followed, including those which are investigational, and have been given a description of the attendant discomforts, risks and benefits to be expected and the appropriate alternate procedures. In signing this consent form, I agree to this method of treatment and I understand that I am free to refuse to participate or to withdraw my consent and discontinue my participation in this study at any time. I understand also that if I have any questions at any time, they will be answered."

The British organizers thought this form was preposterous. As Collins reports it, their American counterparts replied that the goal of this consent form "was to protect the doctor, not to protect the patients, and that if protection of the doctor from lawyers harmed the patient then that was the price one had to pay for doing studies in the USA." Almost twenty years earlier, a Dutch pharmacologist had expressed the opinion that since written consent forms often served only to absolve researchers in case experiments went south, the emphasis on informed consent could also be considered unethical.

Collins watched with interest, and then with concern, as ISIS-2 rolled out. The United States had about 20,000 eligible heart attack patients every month; on average, only twelve were enrolled. British enrollment averaged 180 subjects per month, despite the United Kingdom's much smaller population. The minimal enrollment in the United States was not due to lack of awareness or interest, since American cardiologists were enthusiastic about the study and had participated in its design. Collins believed that the slow enrollment was caused by the lengthy American consent form.

ISIS-2 demonstrated the power of the thrombolytics: Aspirin and streptokinase reduced mortality by a stunning 38 percent. Collins ran the numbers and estimated that if the United States had accrued subjects at the same rate as the United Kingdom, the trial would have been completed—and the treatments available in hospitals everywhere—six months earlier. He concluded that the American consent process was responsible for "directly and predictably causing" about 10,000 "unnecessary deaths."

Collins's published analysis does not provide the calculations to support this number. I worked with Carl Schneider, a professor of law at the University of Michigan, to look at the question more closely. It took almost three years for the trial to accrue enough subjects; if ISIS-2 had been completed earlier, the new therapy would have saved the lives of some heart attack victims. Between 1985 and 1988, 680,000 Americans suffered heart attacks that could have benefited from thrombolytic therapy, and 90,000 died. If ISIS-2 had been completed earlier, the new therapy would have saved the lives of some of those 90,000. But how many?

The actual number is determined by three factors: the amount of time the trial was delayed, the number of patients who could have been saved each month during that time, and the number of cardiologists who were persuaded by the trial's results (some cardiologists were already using the new treatment even before the trial was started). We plugged in these elements and found that the actual cost of the delay was about 6,000 lives. Expanding this calculation internationally suggests that Collins's estimate of 10,000 avoidable deaths is on the money. Each of these elements has a margin of error, but even if all three are dialed down to their minimum, the US death toll is no less than 3,600.

Ethicist David Hunter believes that Schneider and I are wrong. He has not challenged our math, but rather points out, quite correctly, that it takes more time to conduct ethical research. Hunter mentions no examples, but there is little doubt that Chester Southam's research would have slowed if he'd told all of his potential subjects that he wanted to inject them with live cancer cells. Collins's objection, however, was not to obtaining consent but to the needless delay caused by a bloated consent form.

COLLINS'S CRITIQUE CHALLENGES conventional American assumptions of what makes a consent process ethical. American regulators and ethicists often portray comprehensive consent using an overstuffed written form as the only way to meet ethical norms. This view finds few friends on the international scene, where American standards have been widely adopted but are rarely celebrated. Three studies of how heart attack patients react to a proposed experiment, from Sweden, Denmark, and the United States, shed light on the issue.

In 2001, Anders Ågård and colleagues, from Sweden's Lund University, wrote that, "following the adoption of the longer and more formal consent procedure employed in the USA, eligible patients in Sweden and most other European countries must now also receive comprehensive written information and sign a consent form before being included." They designed a study to test reactions to such a form from Swedish doctors and patients.

Many of the doctors believed that their patients were too ill and the forms too long for a genuinely informed consent to be possible, and concluded that the comprehensive consent process was inappropriate. The thirty-one patients who participated were equally hostile. As one said, "I consider it unnecessarily brutal to put a paper under your nose when you don't even know where you are." Several of Ågård's subjects commented that signing the form was done not for their own benefit but for the sake of the doctor. In terms of comprehension, some subjects, questioned later, remembered the study's basic methods: "there were two equally good clot dissolving preparations A and B." Others remembered much less, and two did not know they had been enrolled in a study at all.

One of the subjects said, "I take it for granted that the specialists do what they feel is best for me. I rely on the Swedish health care system." This comment is one sign of how consent practices reflect national characteristics, as medical systems that have established a bond of trust with their patients may have less need of formal, legalistic protections.

Yet even countries with low trust and a need for a more formal consent process face a cognitive ceiling that makes long consent forms counterproductive, because potential subjects everywhere have a limited ability to understand complex information in a medical emergency. Hunter argues that a morally valid consent process must be "substantive." In a best case, this might mean stripping consent forms of their legalese, leaving only the explanation of the experiment itself. But that explanation should not be too long, for ethics does not demand that we give potential subjects information that they do not want and cannot understand.

In 2004, years after Swedish investigators had begun using comprehensive consent forms, Danish requirements remained far simpler. Anne Gammelgaard and colleagues talked with Danish patients who were offered enrollment in a study of heart attack treatment. Legally, the scientists were required to describe the study orally and provide a written form as well, which was just one page long. Afterward, Gammelgaard asked patients, both those who had enrolled and those who had refused, about the process. Most potential subjects wanted to be informed that their doctor planned to enroll them in a study and given a chance to refuse, but they did not find the information sheet useful. Even though it was just one page, many subjects considered it too long. Few read it, and many "found it absurd to expect patients in the acute phase of a heart attack to read and understand a closely written sheet of paper." Gammelgaard still believed that consent should be sought, but that a "morally sound" consent process should be "brief and concise."

Ågård had a suggestion for how this could be done, arguing that it would be "appropriate and generally sufficient" to ask a potential subject, "Are you willing to take part in a study where we are comparing two drugs which are used for the treatment of heart attacks, so that we can find out which one is best? You are free to say no. In that case you will receive standard care. If you want, we can provide you with more information now. Otherwise, we will give you more information about the study later when you feel better."

Back in the United States, ethicist Nancy Kass and her colleagues explored the preferences of patients about studies that compare the effectiveness of two established treatments. They found that patients "wanted to be told about research and have a choice, but were very open to such disclosures being streamlined." Neal Dickert and his colleagues of the Emory School of Medicine interviewed American patients who had recently had a heart attack. Most "preferred to be asked permission prior to trial enrollment." However, Dickert believes that "full informed consent" is an unrealistic goal in view of the press of time and patients' emotional and physical distress. Instead, Dickert's team suggests that the main purpose of the consent process should be seen as giving potential subjects a chance to reject trial participation, replacing informed consent with (sometimes) informed refusal. This approach may be the best way to enroll the willing, leave out the unwilling, and burden neither group with a form they do not want to read.

Hunter thinks we should do more, and argues that we should accept the delays caused by a substantial consent process. He concedes that those delays will cost lives, but he argues that "the number of lives lost that can be legitimately attributed to research ethics regulation is considerably smaller than claimed by Whitney and Schneider ..."

Maybe so. But even if our estimate of 6,000 deaths is cut by half or two-thirds, that still means that adherence to a regulatory ideal, one that provided research subjects with negligible benefit, cost the lives of thousands of people.

IN CHAPTER FOUR, WE SAW how IRB decision making is distorted by a preoccupation with protection of the institution. In chapters five and six, we have seen how an elaborate and ponderous consent process slows research, damages the public welfare, and costs lives. Yet this distorted approach to risk and cumbersome consent process are, it is said, required by ethics.

You may reasonably ask, what role does ethics really play in the oversight of medical research? At this point, midway through our inquiry, we are ready to explore this important question.

# 7

# The Expulsion of Ethics

O N THE STANFORD CAMPUS, a bike will get you anywhere you want to go. The rows of palm trees and sandstone buildings are a visual treat, and the campus engages the senses in other ways, such as the smell of eucalyptus—a little like rosemary but also medicinal, like camphor. It was through these sights and smells that I rode my bike over to the Administrative Panels Office one day in the fall of 1996. I was a new member of the medical IRB, and Donna Adelman, our administrator, gave me my copy of the manual—a three-inch notebook that weighed five pounds. It contained a list of the IRB's members (including James Theodore, our chair), meeting dates, procedures, and much more. She drew my attention to a reprint of the *Belmont Report*, a slim pamphlet that was our ethical guidebook.

The committee needed an ethical guidebook because ethics was our mission. We weren't particularly worried about the science in the proposals that we reviewed, which were generally submitted by Stanford investigators who knew what they were doing. The scientists focused on the science; we made sure that they also protected their subjects from harm and honored their right to a full and appropriate consent.

Sociologist Laura Stark has an inelegant but pithy phrase for the committee's work, writing that IRBs "do ethics." By this she means that IRBs engage in research oversight that uses ethical principles to reach ethical goals. This terminology is mirrored in the name of the academic discipline that studies research oversight: research ethics.

Of course, it was not all ethics, all the time. We did spend considerable effort on paperwork and minutiae, but (sigh) that was one way in which we were like any other medical school committee. And we had to follow the regulations, we all knew that, but that was routine. The promotion and tenure committee, for instance, had to follow university rules. It was the same on the IRB: We had to do what the Code of Federal Regulations required. But my fellow members and I gave the regulations little thought, for the research proposals which we reviewed had all been channeled through an administrative process that tracked the regulations closely. Once it was determined that a proposal *could* be approved under the regulations, it was passed to us to make the yes or no decision that ethics demanded.

ANY SYSTEM OF REGULATION combines papers and humans. The papers contain a set of rules; the humans determine how those rules apply to specific circumstances. Some systems are governmental: The National Park Service has rules about backpacking in the Grand Canyon that are enforced by Park Service employees. Some are private: Hospitals have staff bylaws, administered by a committee of doctors, that determine how a new doctor can get privileges.

General rules can never cover every specific circumstance, so the people who interpret regulations need a tiebreaker for close cases. Government regulators will often attempt to use the public interest as their tiebreaker. When IRBs use ethics as a tiebreaker, it is entirely reasonable to consider their work to be ethics in action. But is that true now? Was it true in the past?

A walk through the system's history shows that IRBs were created before research ethics existed; that the field then emerged and its principles could be applied as a tiebreaker; and that federal pressure then expelled ethics from IRB review, so that today it is absent. Some IRBs may once have "done ethics," but that is rarely if ever the case today.

WHEN NIH DIRECTOR JAMES SHANNON and his colleagues launched the new IRB system in 1966, they did not cast it in ethical terms or ask it to pursue ethical goals. This was a studied decision.

In 1964, Shannon had asked one of his staff, Robert Livingston, to study the relationship between the government and the scientists whose research it funded. Until that time, NIH policy was to give funded scientists liberty to pursue their work as they saw fit. Shannon asked: Should the government attempt to specify how scientists' work could be kept within ethical limits?

Livingston, after considerable study, answered with a ringing "No." Indeed, he wrote, "whatever the NIH might do by way of designing a code or stipulating standards for acceptable clinical research would be likely to inhibit, delay, or distort the carrying out of clinical research." It would be a mistake, he believed, for the NIH to set forth "the definition of ethical boundaries or conditions mandatory for clinical research." Shannon agreed, which is why the earliest IRB regulations were silent about ethics.

Thus, in the early years, an IRB that was uncertain whether or not to approve a protocol would not use ethics as a tiebreaker. Instead, a 1971 federal directive urged members to use their "common sense and sound professional judgment" to serve the public interest.

After the explosion of the Tuskegee syphilis scandal in 1972, the government faced the challenge of identifying what had gone wrong. To many observers, the cause of the scandal was obvious: racism within the Public Health Service, the Centers for Disease Control, and the government more generally. Pioneering civil rights lawyer Fred Gray, who represented many of the victimized men in their later suit against the government, was shocked but not surprised when he learned of the study, for its racial animus was of a piece with the rest of Black life that he knew: "The Study was as racist as segregation in schools . . . as segregation in juries . . . as segregation in the Alabama farm programs, in housing, public accommodations, and in the political process . . ."

But if the problem was racism, that suggested that the legislative response should be race-conscious. That was the last thing the government wanted, and it fought hard to persuade the public that weak review processes were actually at fault. But how should those review processes be strengthened? The solution, Congress decided, was to infuse the regulations with the power of ethics, and it attempted to do just that with the National Research Act of 1974.

The act established an eleven-member commission with a mandate to study the ethics of research. That body, the National Commission for the Protection of Human Subjects of Biomedical and Behavioral Research, gave philosophy a prominent seat at the table. One member was a professor of bioethics, another was a professor of Christian ethics, and all were gathered in pursuit of an ethical vision of research.

The National Commission was empowered to work with the Secretary of Health, Education and Welfare to ensure that newly-drafted IRB regulations would be a force for good ethics. This gave philosophers unprecedented direct input into the shape of regulations. And because the impact of regulations depends not only on what they say but on how they are interpreted, Congress gave the National Commission a second directive: "to identify the basic ethical principles which should underlie the conduct of biomedical and behavioral research." These ethical principles would then be the new tiebreakers as IRBs conducted their deliberations.

IN 1976, THE MEMBERS OF THE COMMISSION spent a long weekend at Belmont House, a conference center of the Smithsonian Institution, to develop a report that would present general ethical principles for research. After the panelists went home, young staff philosopher Tom Beauchamp synthesized their thoughts, added some of his own, and removed much fluff. The result was the *Belmont Report*, released in 1978 and reprinted and distributed widely in the years since.

The *Report* identifies three basic philosophical principles deemed relevant to scientific research. The first, "respect for persons," highlights the value of potential subjects being free to make an autonomous choice about whether or not to enroll in research. The second, "beneficence," emphasizes the duty of the IRB to protect the welfare of research subjects. The third principle is "justice," which IRBs satisfy by ensuring that the choice of subjects is equitable. By the *Report*'s account, no one of these principles is dominant. Yet it is undeniable that philosophers have a weakness for personal autonomy; Beauchamp himself calls it "the most important principle." The primary manifestation of this principle in the field of research ethics is subject choice through the process of consent.

The *Report*'s approach assumes that research entails some burden or risk, since in their absence consent is of less moral weight. It also assumes that individuals' goals may go beyond advancing their personal interests; otherwise, subject welfare (beneficence) would unconditionally trump autonomy, and the discussion would go no further.

The *Report* was soon hailed as the ultimate guide to ethics in American research, a position it retains today. Other statements of ethical principles, most notably the Helsinki Declaration, are more influential internationally, but in the American context, the *Report* is generally considered definitive. Ethicists Franklin Miller and Jonathan Kimmelman call it a "landmark." Lydia Furman, a pediatrician who serves on the IRB of Cleveland's Rainbow Babies and Children's Hospital, calls the *Report* "a breath-taking document" that is a "must read."

Not every ethicist agrees. Although the *Report* claims that its principles are "comprehensive," it omits the moral value of research that can save lives. It also assumes that research ethics should focus on the experimental subject as an isolated individual, disregarding the rich human relationships that form our lives and our society. Feminist and communitarian schools of ethics recognize the value of these relationships, and the corollary that IRB oversight should not keep potential subjects from accepting a measure of risk in the hope of helping others.

Yet the gap caused by the absence of a larger moral vision can be reduced by the *Report*'s emphasis on autonomy. Through autonomous choice, research subjects can put their ideals into practice. If they value family, they can choose to participate in research that will do them no good but might benefit their children. If they value society, research participation gives them a chance to help strangers whom they will never meet.

This emphasis on autonomy is why the *Report* represents a significant advance from Hans Jonas's 1969 essay. His vision, with its single-minded focus on protecting subjects from harm, was black and white. The *Report* presents a dappled ethical landscape in which autonomy, welfare, and justice all deserve our respect.

Although Jonas and the *Report* agree on the importance of autonomy, they disagree on how it should be seen in the context of research. For Jonas, the "invitation" to enroll in research is an offer that cannot be refused and that crushes the autonomy of potential subjects. The *Report* takes the opposite view. It begins by assuming that potential subjects are not coerced—they have a genuine choice to make. Genuine choices are dear to philosophers' hearts, for they permit people to express their moral convictions even at some cost. In this view, consent is no longer merely a shield protecting researchers and their institutions from liability. It is an ethically valuable opportunity for potential subjects to use their autonomy to make morally significant decisions.

Willing research subjects thus express their autonomy by accepting some risks in pursuit of their values and preferences. Society relies on the consent process to present those hazards fairly so that subjects may choose wisely. If subjects have a right to make autonomous choices, then IRBs, which have a duty to promote subjects' rights and welfare, are responsible for enabling those choices whenever possible. More specifically, if autonomous individuals wish to participate in research—assuming that they are not part of a group that has been unjustly selected to participate, that the hazards of the research are not severe, and that they are adequately informed about those hazards—they have a positive right to enroll, and the IRB should not overprotect them by blocking that right.

As to the question of what this means in terms of routine IRB deliberations, the *Report* states that it intends to "guide the resolution of ethical problems" in research by discussing the implications for each case of autonomy, beneficence, and justice. Because it stipulates that no single principle should automatically prevail, the *Report* provides IRBs with more flexibility than guidance. But as philosophers imagine how IRBs make decisions, this indeterminacy is a feature, not a bug. No single answer will fit every situation, and so each IRB must make its own best choices. In the common and important case of an experiment with risks that are small but real, the welfare of potential subjects is in tension with their autonomous right to accept a measure of risk. Some IRBs may decide that, in a particular case, protecting subjects' autonomy justifies allowing some additional risk. Other IRBs, reviewing the same protocol, will deem the risk too great. The latter committee thus deprives potential subjects of their autonomy, but only after careful thought.

The *Report* completed the moral framework that Congress had envisioned. The regulations embodied generalized ethics, and the

*Report* showed IRBs how to apply ethics in specific situations. With ethics defined and prescribed at both levels—the regulations and their application—its place seemed secure.

Federal officials, who had once collectively given ethics the cold shoulder, soon saw the advantages of being seen as ethical champions. To this day, OHRP links to the *Belmont Report* on its home page, and it reinforces the importance of ethics by requiring each federally-funded institution to promise that it will be "guided" by the *Report* or a similar statement of principles. But ethics is nowhere to be found in OHRP's actual operations. It was Jerry Menikoff himself who pointed out, before he became the agency's director, that OHRP's comprehensive list of potential violations to be looked for in an audit mentions the *Report* not once.

On the scholarly side, the *Report* became part of the booming field of academic research ethics. Henry Beecher had stood almost alone with his 1966 exposé. In time there were professorships, institutes, conferences, and journals that published a swelling volume of commentary on every aspect of how the relationship between scientist and subject should be overseen: how much risk is generally acceptable, for instance, and when it was appropriate to waive consent, and on which research, if any, should be exempt from review, and on the ways in which the regulations provide too little or too much protection for children, or the dying.

ONE AREA ATTRACTED NOTABLY LITTLE INTEREST: the extent to which IRBs actually used the *Belmont Report* principles in their deliberations. To put it more bluntly, did IRBs really embrace ethics as the tiebreaker?

Some have certainly tried. In 2003, James Anderson and his colleagues at the University of Nebraska described how their discussion of an early-phase drug study could be mapped to the *Belmont* principles of respect for persons, beneficence, and justice.

They did not say that their IRB used this process routinely. No doubt other IRBs attempted to use ethical principles in their work, but how well they succeeded and how long they hewed to ethics as a tiebreaker is not known.

I served on the Stanford IRB twenty years after the *Report* was published, surely enough time for its concepts to be fully assimilated into our work. Yet I don't remember our IRB ever debating the ethical aspects of a protocol or mentioning autonomy, beneficence, or justice. Our members, mostly scientists, were more at home analyzing a trial's risk, enrollment, and randomization. We believed we were doing ethical work, but we did not actually use the *Report*'s principles as its authors had hoped. Were other IRBs of the era using ethics more consciously? We don't know, for the topic had attracted no scholarly attention.

This changed in the early 2000s, just after Ellis's death penalty era was drawing to a close. Sociology graduate student Laura Stark persuaded three IRBs to allow her to sit in on their meetings while making audio recordings and jotting down field notes. She supplemented this research with interviews with IRB chairs at major research universities. She published her findings first as her dissertation and then, with additional historical material, in the book *Behind Closed Doors: IRBs and the Making of Ethical Research*. Her observations are particularly valuable because her broad sample ensures that they do not reflect merely a single idiosyncratic committee.

Stark's findings are as useful for what she did not observe as for what she did. In her review of dozens of IRB meetings, she never reports a committee expressing concern for prospective subjects' freedom to choose to participate in research; instead, boards routinely restrict subject choice in their attempt to ensure subject safety. They ignore autonomy and focus on protecting prospective subjects, and themselves, from risk. Stark never reports an IRB using the *Belmont* principles to engage in ethical debate. In fact, as she observes, "abstract formulas and general principles like those described in the foundational texts of bioethics such as the *Belmont Report* cannot actually be used, in any practical sense, to make decisions ..."

Nor did Gary Ellis launch the death penalties of 1998–2001 through the application of ethical principles. Instead, he cited what he saw as serious violations of the regulations to justify turning off federal research funding. This forced institutions to realize that their IRBs were now a vital protection against federal sanctions.

It also forced ethics out. This expulsion of ethics was a result of the severity of federal punishment and, even more, of its unpredictability. In this new regulatory state of affairs, any sensible IRB, given a decision that could go either way, would use the avoidance of federal sanctions as a tiebreaker. As sociologist Caroline Bledsoe and her colleagues write, "the IRB's over-riding goal is clear: to avoid the enormous risk to the institution of being found in noncompliance by OHRP."

This risk cannot be eliminated, in part because of the limitless variation of research topics and methods. IRBs must often address a protocol that is simply a poor fit for the regulations. This would not ordinarily be a problem, for dealing with the ambiguous and the unusual is the specific challenge for which deliberative bodies like IRBs are most valuable. But that requires them to be able to use their best judgment and have faith that OHRP will not later arbitrarily decide that they were wrong. IRBs that take the obvious step of requesting OHRP assistance have not generally found the agency to be useful.

Amid all of this confusion, many IRBs have worked with scientists in their institutions to work out more-or-less standard approaches to common research questions. This allows much research to proceed without undue delay, although little variation may be allowed. But even this is an unreliable safeguard. A major study of the use of oxygen in very premature babies, approved under what seemed to be standard terms by twenty-three IRBs, was begun in 2005 and completed in 2009. Four years later, OHRP launched a slashing attack on the investigators and the scientists involved. We will look at this case in detail in chapter nine.

All regulations provide flexibility in any given circumstance, so IRBs once had a measure of freedom to follow ethical principles. Now, what former OHRP director Greg Koski calls "reactive hyperprotectionism" rules the day. Whenever the regulations leave room for discretion, IRBs must seek the choice that best protects them from federal sanctions. As a result, the IRB's main question becomes, "Does this study pose risk—any risk of any kind?" Risk, which was once (under the label "beneficence") one ethical principle among three, has swallowed the others.

Furthermore, the nature of risk itself has changed. Once it related to risk to subjects; now risk to the institution is the primary fear that drives decisions. The idea that one of the committee's major responsibilities is to protect the institution itself, which would have made no sense thirty years ago, is now commonplace. Sociologist Sarah Babb found that IRB administrators and chairs see protecting their institution as an "openly acknowledged . . . and . . . common-sense part of their jobs."

I know that some IRB members, reading this section, will object that they have served on their IRB for years and never heard a whisper of this supposed federal menace. "In our institution," they will say, "we make our own decisions, and while IRBs at other institutions may be running scared, we are not." I understand this reaction because it's exactly what I would have said, based on my own experience.

This issue highlights the very different positions occupied by the IRB's members, chair, and administrative staff. While the committee chair and staff must keep federal attitudes in mind at all times, members are generally sheltered from this unpleasant reality. They come to meetings and share their opinions about specific proposals, vote, and go home. The proposals they see have been processed by the IRB staff, usually under the combined supervision of the senior IRB administrator and the IRB chair.

So you could be a long-standing IRB member and never realize how much the chair of your committee, and the administrative staff, are motivated by fear of federal sanctions. You might never realize that the university's legal department is concerned that a particular protocol might be problematic. And James Theodore, our chair at Stanford, was so skillful that I am sure he could quietly persuade the committee to table a problematic protocol, perhaps for a long time, without ever hinting that he was under any kind of pressure.

This is not just a matter of the chair not sharing an unpleasant reality with the members. To mention the possibility of federal punishment might invite members to contemplate the fact that they could be singled out as targets, as when OHRP action provokes a lawsuit that names the individual IRB members as defendants (as we'll discuss, this was another element in the study of oxygen in premature babies). At a time when IRBs struggle to attract and retain members, a wise chair will not dwell on committee members' legal liability

I present these observations about how IRBs view federal supervision as common in the IRB world, not universal. Some IRB chairs perceive OHRP as helpful, or at least as moving in that direction. When Columbia psychiatrist and ethics researcher Robert Klitzman surveyed IRB chairs about their relationship with their federal overseers, one told him that OHRP was becoming "much more approachable. Recently, I e-mailed them a couple of questions. It took them a month and a half to get back to me." This was an improvement, since in the past the agency would never answer in writing. (An agency staff member would sometimes call, which provided little support if the IRB's actions were later challenged.)

In its most significant recommendation, the *Belmont Report* asked IRBs to allow potential subjects to make real choices. Today's ethics are nominally unchanged, but the IRB system's daily operations reduce the choices on offer. The theory of ethics thus remains everywhere in the system's thought and writing, while its practice has been squeezed out. We are left with a system of regulation that imagines it is a system of ethics.

This is how ethics was welcomed into the heart of the IRB system and then ejected from it. All that remains is the husk—the talk of ethical principles, the references to Tuskegee and those reprints of the *Belmont Report*. The ever-more-elaborate bureaucracy is fully in charge. In sociologist Carol Heimer's summary, "Ritualistic legally based regulation of research ethics makes research less ethical."

Ethics is only one of the IRB system's many interlocking parts. From the point of view of the scientists, its main feature is how it influences their ability to generate new evidence to support better medical care. And on this point—how research should be conducted, and how doctors use that evidence to care for patients—scientists and OHRP have fundamentally different positions, as we will see in the next chapter.

# 8

# Evidence

M R. M'S ORDINARY MORNING took a turn for the worse when he developed severe chest pain. He went to Manhattan's Beth Israel Hospital, where doctors diagnosed a heart attack. As was common in 1979, they started him on oxygen, morphine, and an intravenous drip of lidocaine to stabilize his heart rhythm. I was a medical student, and soon after Mr. M was settled in his bed in the coronary care unit, I pulled up a chair, introduced myself, and asked if he would be willing to talk with me for a few minutes.

Mr. M. was happy to tell me about his eventful morning, but before long he began struggling, first to find words and then to keep his eyes open. I watched, appalled, as he slowly lost consciousness. His head slumped forward on his chest, and when I shook him he did not respond. I wondered if this was the result of an arrhythmia, but when I felt his pulse it was strong and steady.

I sat by his side, my mind racing and my panic mounting, until inspiration struck: Get help! I ran to the nurses' station and said there was a problem with Mr. M. The nurse strode past me into the room. He knew something that I did not—that lidocaine is a general anesthetic, which could be used as such in the operating room, as well as a cardiac rhythm stabilizer—and he turned the drip off.

Before long, Mr. M's eyes fluttered open, he looked around, and soon he was wide awake again. He knew something had gone wrong, and I explained that he had been getting too much medication but that he was going to be fine now. He looked me in the eyes as he squeezed my hand gratefully. "You saved my life," he said.

There remained the problem of his heart attack. Mr. M's chest pain was caused by a clot in the arteries that led to his heart. So far, he was doing okay, but there was always the danger that in a matter of hours enough muscle would die that his heart would stop. This was before ISIS-2 proved the value of clot-dissolving drugs, and doctors had little or nothing to dissolve or bypass the clot and prevent death. Even patients who lost only a small amount of heart muscle were vulnerable to death from a second process, the disruption of the heart's normal rhythm.

A single heartbeat is a complex event. The heart's upper chambers, the atria, swell as they receive blood from the body and the lungs. When the atria contract, they force that blood into the lower chambers, the ventricles. Ventricular contractions send that blood to the lungs and the body. Thus, there are two contractions for each heartbeat, the first in the atria and the second in the ventricles. Timing matters, for the heart works best when the ventricles have just enough time to fill before they contract. When the ventricles contract too early—an event known as a premature ventricular contraction, or PVC—they do not have time to fill, so little blood is pumped, and the effort is wasted.

An occasional PVC is of no importance. In patients with heart attacks, however, a brief run of PVCs, even when it causes no symptoms, sometimes deteriorates into more dangerous rhythms. Prolonged runs of PVCs can reduce the flow of blood sufficiently to cause lightheadedness or fainting. They also pose a danger of degenerating into a lethal rhythm such as ventricular fibrillation.

Even a small clot, if it damages the nerves that direct the heart's rhythmic pumping, can thus lead to sudden death.

Every coronary care unit has a bank of monitors at its central station that allows nurses to see abnormal heart rhythms in real time. A patient's rhythm might exhibit, for instance, asymptomatic short runs of PVCs—step one—before degenerating into symptomatic longer runs—step two—and ultimately into ventricular fibrillation.

In the 1970s, some cardiologists began to use antiarrhythmic drugs for short asymptomatic runs, as a way to keep them from deteriorating further. These patients remained at risk of sudden cardiac death even after hospital discharge, and so the drugs were continued at home. By the time I met Mr. M, cardiologists had given these medications to millions of patients in hopes of preventing sudden death.

These antiarrhythmics were never approved by the Food and Drug Administration for that purpose—they were being prescribed off-label—but that is a normal and inevitable state of affairs. It would be impossible for the agency to evaluate the fitness of every new drug for every possible purpose. Thus, although there had been no randomized trials of the effect of antiarrhythmics on patients with mild rhythm disturbances, this use had become common practice. In this regard, it was like many other treatments that we could call either "standard" or "habitual" but are not grounded on direct experimental evidence.

Several years passed after my encounter with Mr. M. By the mid-1980s, the merits of antiarrhythmics for patients like him had become a matter of some debate. The FDA had approved medications, including encainide and flecainide, for use in arrhythmias that cause palpitations, dizziness, or loss of consciousness (flecainide is still used for that purpose today). There was no proof that patients like Mr. M, with milder rhythm problems, would be helped by these drugs, yet most cardiologists prescribed them routinely in the belief that they provided an important benefit.

This conviction was based on intuition, not on evidence, and it had apparent limitations. Not every patient with short runs of PVCs goes on to develop more worrisome patterns; the antiarrhythmics had never been evaluated for this particular use; and any medication that can improve the heart's rhythm can also worsen it. But there is never enough evidence to guide every medical decision. Intuition is thus the essential putty that fills the gap between generalizable evidence and the unique factors of each individual patient's situation. To put it more crassly, when doctors don't know, they guess.

The antiarrhythmic boosters might have been offended if their beliefs were described as guesswork. From their point of view, the argument for this use of the drugs could be distilled into a simple syllogism: Abnormal heart rhythms cause sudden death. Antiarrhythmics prevent those abnormal rhythms. Therefore, antiarrhythmics save lives.

How many lives? Well, about three patients per hundred experience a fatal arrhythmia, and heart attacks are common. This suggested that the treatment was saving thousands of lives every year.

The intuitions of other cardiologists pointed in just the opposite direction, and their patients were denied the benefits, or spared the harms, of antiarrhythmic treatment. In 1979, these skeptics, who were definitely in the minority, refused to prescribe lidocaine, and in 1985, they refused to prescribe the newer antiarrhythmics encainide and flecainide.

With the cardiologists at loggerheads, it was time for a trial. Such a study would need thousands of patients and would be beyond the resources of any individual hospital. Fortunately, the National Heart, Lung, and Blood Institute took the lead, and in 1986, with the cooperation of scientists at ten medical centers, NHLBI launched the Cardiac Arrhythmia Suppression Trial. Subjects in CAST, who had all had heart attacks and experienced worrisome abnormal rhythms, were assigned at random to receive flecainide, encainide, or a placebo.

CAST was planned to last for three years, but after two years the emerging data forced it to an early halt. About 750 subjects had been given a placebo, and a similar number received one of the active drugs. Total deaths were 26 in the placebo group and 63 in those given an active drug. The antiarrhythmics—widely used on the basis of physicians' intuition—were killing patients in large numbers.

After the disastrous results of routine practice were exposed, the flaws of the justifying syllogism became obvious. Flecainide and encainide do not always reduce arrhythmias; they can promote them as well. If doctors knew in advance which three heart attack patients out of any given 100 would suffer death by arrhythmia, they could treat those three patients, and lives would certainly be saved. But treating everybody meant that 97 patients who were never going to develop serious problems also received the drugs, and were thus exposed to the drugs' tendency to create dangerous rhythms. As a result, the number of patients who were saved by the drugs was dwarfed by the number who were killed.

Nobody has calculated how many patients died unnecessarily in the years when this treatment was in common practice, but we can sketch in the numbers. In CAST, there were 37 additional deaths in the 750 subjects who received antiarrhythmics, for a rate of 5 percent. There were 760,000 heart attacks in the United State in 1987. If 10 percent of this population was treated with the antiarrhythmics and 5 percent of those patients died as a result, these drugs were killing about 3,800 people every year. This is an educated guess. But whatever the exact number, the basic fact is clear: Doctors were killing patients because they did not have the evidence they needed.

This is tragic, but it is not surprising, for the reality of medical practice is that there is never enough evidence to go around. In the absence of relevant data, doctors must rely on intuition, which draws on their training, experience, and knowledge of similar cases. Off-label prescribing continues because we have no choice.

Here's another example. There have been no formal studies of the effectiveness of many antibiotics in many infections for which they are in everyday use. With more than a hundred antibiotics and thousands of infectious diseases, to test every drug for every infection would require innumerable expensive studies. The studies that are conducted are usually those where evidence might have the greatest impact. Each study of this kind helps doctors as they strive to practice what is called evidence-based medicine.

The common cold illustrates this struggle. Early in the antibiotic era, there were high hopes that it could be cured. Colds are not life-threatening, but their cumulative economic toll is substantial, as would be the profits of any company that marketed a cure. Study after study, however, showed that antibiotics make not a particle of difference in ordinary colds or in any other viral infection. This axiom remained true during the recent Covid pandemic—although this evidence didn't stop Covid patients from demanding antibiotics like azithromycin, just as it doesn't stop them from demanding antibiotics when they get a cold.

Popular treatments that are eventually debunked extend beyond the pharmacy. In 2002, Bruce Moseley, an orthopedic surgeon at Baylor College of Medicine, published a study of arthroscopic surgery for knee arthritis, a procedure that American doctors then performed 650,000 times each year. Mosely found that sham surgery, in which the patient is sedated and a shallow incision made in the skin, works just as well as a real operation. Most orthopedic surgeons did not celebrate this important finding, since it meant that they would be giving up fees totaling billions of dollars.

Medical evidence, then, is more rare—and more valuable—than most laypeople recognize.

You may feel a yawn coming on as you read about the virtues of evidence. Surely the importance of evidence in medicine is obvious. But while everyone celebrates evidence-based medicine, in practice evidence is scorned wherever it goes. When evidence stands between them and an antibiotic they want for their cold, patients become demanding. When evidence shows that a surgery that pays the rent is useless, doctors become defensive. And when scientists conduct the research that would replace guesswork with evidence, OHRP investigates.

AN OHRP INVESTIGATION THAT ILLUSTRATES THIS POINT related to research in intensive care units, where treatment uncertainty is omnipresent. There is, for instance, the question of how to treat sepsis, a condition that claims the lives of 250,000 Americans a year. Sepsis is the body's reaction to a severe infection, and it can cause a precipitous drop in blood pressure. This fall in pressure can be treated in two ways: Some doctors give medication, others give extra fluid. Both work, at least some of the time, and nobody is sure which is better. Doctors sometimes justify their preference by saying, "in my experience, this is the most successful approach," but "successful" is a stretch for a condition in which a death rate of one-third is pretty good.

This is where PETAL comes in.

PETAL (Prevention & Early Treatment of Acute Lung Injury) is a consortium that includes leading specialists in pulmonary and intensive care. It is funded by NHLBI to conduct major trials, often of treatments that are used every day but whose actual benefit or harm has never been carefully studied. The importance of these trials was highlighted during the early days of the Covid pandemic.

In 2020, Covid's attack on the lungs was so savage that ventilator shortages became major news—how many were needed as things got worse, the efforts of governors to secure enough for their states, and the work in design and manufacturing to build machines rapidly. People were interested in how ventilators worked, and the media obliged. A *New York Times* reporter, describing a very sick 49-year-old man, wrote about doctors who use ventilator settings that are "precisely calibrated and continually adjusted." At another point in the article, though, she commented that the doctors were riding "roller-coasters of trial and error" involving the "risks of uncertain treatments." The first of these statements emphasizes expertise, the second uncertainty, but there is no conflict between them. The dials on a ventilator do allow the doctor to set the precise depth of each breath, the level of oxygen, and much more. But doctors disagree on what those settings should be, and patients often die no matter what settings are used.

Severe Covid infection often begins as pneumonia. As it worsens, pneumonia can progress to lung failure, which in this context is called acute respiratory distress syndrome. Pneumonia is not the only cause of ARDS; it can also be triggered by burns, trauma, and the inhalation of toxins. Across the board, about half of all patients with ARDS die. Sometimes the underlying cause can be treated, but there is no treatment to reverse ARDS itself—all the doctors can do is to try to keep you alive on a ventilator until your lungs recover.

The PETAL consortium, which focuses on conditions like ARDS, has a distinguished heritage. ARDSNet, its predecessor, was founded in 1994 to tackle the treatment conundrums that abound in the ICU. For patients whose lungs have failed, for instance, there were two schools of thought on how deep each ventilator breath should be. Some doctors favored forced deep breaths, which deliver ample oxygen and keep carbon dioxide from accumulating. Others pointed out that forceful ventilator settings can rupture delicate lung tissue, and they used shallow breaths, which were gentler on the lungs. This was no panacea, since these shallow breaths led to terrible air hunger, and patients would rip out the endotracheal tube in an attempt to breathe more freely. Sedation eased their air hunger but carried hazards of its own. Thus, the treatment that patients received varied based on which doctor happened to be providing their care, since there was no evidence as to which method was better.

In 1996, the ARDSNet investigators began a depth-of-breath study in ten major medical centers. For each patient enrolled, a treating pulmonologist called a central phone number with an interactive voice system, which randomly assigned that patient to receive either shallow or deep breaths. The results, published in

2000, showed much better survival in the group given shallow breaths. Out of about 430 patients in each arm of the study, there were far more fatalities in the deep breath group—172 versus 133. It is worth noting that even the deep breath group's mortality, at 40 percent, was good in comparison to the usual outcomes. This may be a reflection of the benefit of being in a carefully-designed trial in which every detail of care is specified by a group of experts.

The depth-of-breath study was a major breakthrough in the treatment of ARDS. For the first time, pulmonologists had solid evidence about how they should set their ventilators, and these more shallow breaths (six ml/kg was the winner) became the nationwide standard, saving thousands of lives every year.

The ARDSNet studies were conducted sequentially: One trial could begin only after the prior one had finished. With shallow breaths the new standard, ARDSNet scientists next investigated the difficult question of how much fluid patients on ventilators should receive. Once again, there were two competing schools of thought. Some doctors favored providing relatively generous amounts of fluid, which increased the amount of blood circulating to the brain and kidneys (on rounds, we would call these patients "wet"). But fluids had a disadvantage as well, for wet patients have soggy, inefficient lungs. So the relative merits of keeping patients relatively wet or dry had been the source of much debate. As Dr. Herbert P. Wiedemann said, "We have been flying blind."

In an effort to settle this debate, in June 2000 ARDSNet launched the Fluid and Catheters Treatment Trial. FACTT was the product of an extensive developmental process in which dozens of experts considered every detail of the study's method. Those involved included a steering committee, the NHLBI project scientists, a clinical coordinating center, a protocol-review committee, a data safety and monitoring board, a consultant from ARDSNet, and representatives from 20 clinical centers—a total of about 160 experienced scientists. If you are curious about the methods used to study a complex question, you can browse through the protocol, which can be downloaded (it's not behind a firewall) at https://www.nejm.org/doi/suppl/10.1056/NEJMoa061895/suppl_f ile/nejmoa061895sa1.pdf. FACTT rolled out in June 2000, with a planned enrollment of 1,000 patients. Within two years, 418 patients had joined.

Then OHRP stepped in, halting the study and launching an investigation. The agency was responding to concerns raised by two critical-care doctors who were not involved in ARDSNet, and who felt that the depth-of-breath study—not the current study, but the previous one—had been unethical. These skeptics believed that ventilators should be set to give neither shallow nor deep breaths, but breaths in the middle range. Since none of the patients in the depth-of-breath study had received these moderate breaths, they believed that the trial was unethical and had harmed its subjects. And since its conclusions were the foundation for the FACTT trial, in their opinion that follow-up study was also unsafe and therefore unethical.

OHRP chose to believe the skeptics and discount the scores of scientists who had designed the trial. The study was put on hold. Every center was notified: No subjects were to be enrolled until further notice. The trial's clock stopped on July 25, 2002.

There is nothing wrong with raising doubts about an experiment that is underway, since even the most distinguished scientists can be in error. But in ordinary circumstances, a trial's science and ethics can be evaluated without stopping it in its tracks. OHRP never explained why it thought the peril was so great that the trial had to be halted.

Disputes over study design can sometimes be resolved by discussion among the parties. When that fails, as it did in this case, the best solution is often to ask for help from an independent panel, formed for this specific purpose. In this instance, NHLBI, after requesting input from all interested parties (including OHRP), appointed an independent five-person panel to review the study's design and ethics. The panel met on August 30, just over a month after FACTT had been halted.

Panels can deadlock and leave matters as muddled as before, but this one was unanimous. It declared the study to be safe, scientifically sound, and likely to yield important results. The panel's chair, Mitchell M. Levy of Brown University, wrote, "The logic for almost every issue is based on the assertion that there is, or could be, an identifiable 'current standard of care' or 'routine care' for either mechanical ventilation in ARDS or fluid management. The literature is fraught with debate and controversy about the standard of care for both of these treatment modalities. . . . It is this ongoing variation that lends such importance to both the . . . trials." Another panel member urged OHRP to let the "ARDSNet investigators proceed with their landmark, world class investigations."

With this resounding vote of confidence, the investigators waited hopefully for a green light from OHRP. Instead, the agency dug in its heels. On October 7, when the study had been on hold for more than two months, it sent the investigators a 29-page letter. It faulted them for not providing at least some of the subjects in the depth-of-breath study with "individualized mechanical ventilation management . . . based upon consideration of a number of complex clinical factors unique to each subject."

The attitude toward medical treatment implied in this statement reflects one that many people have. Jerry Menikoff, before he became head of OHRP, wrote, "how would you feel if your doctor suggested— not as part of a research study—that he pick the treatment you get by flipping a coin? Very few of us, as patients, would accept this type of behavior. In real life, we rarely make decisions by flipping a coin. When there are complicated decisions to make, we do the best we can using the limited information we have. Ignoring relevant information, even if it is not very reliable, is rarely an ideal way to achieve a specific goal."

Menikoff's comment echoes the feelings of patients who prefer doctors who claim to know which treatment is best. Like Menikoff, these patients are disappointed when a doctor says, "There are a couple of treatment choices for your situation, and nobody knows which is best." They prefer a doctor who takes whatever medical knowledge is available, adds in experience and intuition, and makes a single confident recommendation. These patients may know that another doctor might make the opposite recommendation, but *this* doctor has examined them, knows their story, and has gained their trust.

This divide extends to the realm of research. Some patients who are offered enrollment in a clinical trial readily understand that there are two treatments, each with its pros and cons, and that the study is intended to figure out which is better. These patients may not relish this uncertainty, and the fact that which treatment they get will be determined more or less by the flip of a coin, but they realize that sometimes life is like that.

Other patients have just the opposite reaction. They are uncomfortable with uncertainty and are convinced that the doctor just needs to dig deeper to find the better treatment. In this context, "digging deeper" amounts to using the doctor's intuition, which is sometimes a mental flip of the coin, but if the doctor doesn't say that out loud, the patient is satisfied.

And here's the extraordinary thing. When OHRP puts itself in the position of someone who is considering enrolling in a trial, it doesn't act like a patient who knows that medicine is an uncertain science. Even though Menikoff is a doctor who has published original research, and the agency boasts ample staff with expertise in biomedical science, OHRP still operates on the assumption that doctors' intuition may be relied on even when the evidence is a muddle.

ARDSNET RECOGNIZED THAT THE MANAGEMENT of patients on ventilators was an area of substantial uncertainty. Each of the ARDSNet doctors had their own customary ventilator and fluid settings, but they knew that other doctors would choose differently.

When a patient's oxygen level fell, they knew that deeper breaths or less fluid would bring it up. But even if the oxygen level improved, the ventilator change might cause lung injury and the reduced fluid might cause kidney failure. Or not. There are too many moving parts to be confident about the role of any given element, and too many patients die to feel confident in current practice. This is why the pulmonologists were so happy to know that one element, the depth of ventilator breaths, worked reasonably well at six ml/kg.

With its letter of October 7, OHRP put the stamp of federal authority on the belief that, even when the evidence is complex and unclear, a compassionate doctor, dedicated to the patient's well-being, will somehow have the experience and intuition to provide superior care. That is magical thinking, and it would have destructive results.

The investigators were too politically astute to accuse OHRP of magical thinking. They did push back, pointing out that routine care, no matter how individualized, was exactly the quagmire they were trying to rescue patients from, and the data drawn from adding a group of subjects who were given routine care would be "almost entirely meaningless." Although OHRP demanded further data, and justification, and analysis, it could not escape the fact that the independent panel, which included members approved in advance by OHRP, had resoundingly supported the study just as it was.

The agency was still not ready to admit its error, and so it created another panel, whose members it alone selected. On July 3, 2003, almost a year after it had halted the study, the agency announced the recommendations of its consultant panel, whose eight members included specialists in research ethics, bioethics, critical care, and statistics. The agency was forced to concede that "almost all of the consultants engaged by OHRP opined that risks to subjects participating in the FACTT trial were minimized and reasonable in relation to anticipated benefits to the subjects and the importance of the knowledge that was expected to result." The agency provided little additional information about their views, and if any of the consultants were critical of OHRP's decisions to stop the trial, and to continue that halt despite the findings of the first panel, those comments were not made public. In case someone wanted to contact the panelists directly, the agency took the additional safeguard of not releasing their names.

OHRP never admitted that it should not have suspended the study, but one fact was undeniable: Its investigation had delayed the study's completion by a year. When it was finally completed, it showed that patients kept relatively dry could be taken off their ventilators faster—another finding that would save lives.

Because it is costly and complex to do research that combines investigators and patients at multiple institutions, the one-year delay of FACTT had a domino effect on subsequent research. We cannot know if the delays to follow-up research would now amount to more or less than one year. We do know that in 2013, ARDSNet was dissolved and reformulated as PETAL, which conducted a trial of sedation in ARDS, followed by a trial of vitamin D3. Both studies were completed and finished. In 2018, it launched a study of the treatment of sepsis. By early 2020, it had enrolled 1,300 patients out of a planned maximum of 2,300.

Then the pandemic struck, and while sepsis continued to kill as many patients as ever, the critical care community devoted its resources to the struggle against Covid. If the sepsis study had been finished before the pandemic shut it down, doctors could have used its results during the Covid crisis.

I do not know why OHRP acted as it did. Perhaps it failed to understand the role of evidence in medical care. Perhaps it wanted to act tough. Perhaps there was some other cause. In any case, it was not the agency's most troubling action. For that, we need to turn to a recent study of how best to care for premature babies, an issue that has been a matter of ongoing investigation since the 1940s. It is a fascinating if disturbing story that will take us from before the IRB system's creation to its contemporary struggles.

# 9

# Babies

I N FEBRUARY, 1941, DR. STEWART CLIFFORD, a Boston
pediatrician, made a routine house call to check on a three-
month-old girl. The baby had been born prematurely, did well with
intensive care, and was sent home without problems. This was a
remarkable achievement, since not that long before she would
probably have died.

A healthy pregnancy lasts about 40 weeks from the time of the
mother's last menstrual period. Premature babies, those born before
37 weeks, struggle to breathe, feed, and keep warm, and efforts to
meet their special needs have a long history. The first attempt to use
oxygen in their care took place in 1780, the first warming devices date
to 1835, and incubators were in occasional use from 1896 onward.
Hospitals gradually improved their programs for the care of these
infants, and by 1940, incubators, which could provide warmth and
supplemental oxygen, were in widespread use.

The family thought their baby was fine, but Clifford was shocked
to discover something abnormal in her eyes, something that blocked
the passage of light. Babies are sometimes born with cataracts, which

can be surgically removed, so Clifford asked for a consultation from Fredrick Verhoeff, a Harvard faculty member and eminent ophthalmologist. Verhoeff was stumped—he had never seen a condition quite like this—but he was clear on one point. No treatment was possible. The baby was blind.

Within a week, Clifford was called by another family, whose son had also been born prematurely and responded well to intensive care. He was now seven months old, and his family had noticed that his eyes were moving strangely. Verhoeff consulted again, and he confirmed that this was another case of the condition, which had never before been described. Within a year, fifteen cases had been diagnosed. The problem was not confined to Boston, since once the Harvard doctors reported their findings, cases were soon diagnosed all over the world. The number of blind babies mushroomed first into the hundreds and then into the thousands.

Careful observation showed how the condition progressed. The afflicted babies were all premature, required intensive care, and had normal eyes at birth. The first abnormality noted was a subtle enlargement of the blood vessels in the retina; as the vessels enlarged further, the retina became elevated. At this stage, the disease sometimes improved on its own. When it worsened, the retina gradually became, as the ophthalmologists say, detached, meaning that it had torn loose from its moorings. Once this process was advanced, the eyes moved abnormally, the pupils became white, and the child would be either severely visually impaired or completely blind. Because of the retinal damage, the disease eventually came to be called retinopathy of prematurity. The cause was unknown, the course was unpredictable, and the attempts at treatment during the 1940s—one infant in the Harvard series had been subjected to X-ray therapy, for instance, and another underwent eye surgery—had failed.

The few clues as to the cause of this terrible problem were cryptic. It soon became apparent, for instance, that retinopathy was more common in a certain type of hospital. Disease is ordinarily eager to strike in humble institutions, so once retinopathy had been diagnosed at Harvard, it should have been found in equal or greater numbers in every facility with a premature nursery. Just to the contrary, infants treated in less prestigious hospitals were largely spared. As neonatologist William Silverman described the riddle, the disease "seemed to occur most frequently in infants reared in premature infant nurseries with the most highly organized and advanced programs for care!" The exclamation point is Silverman's. International comparisons showed a similar pattern, with the disease more common in developed than developing countries.

Silverman, who ran the premature nursery at Babies Hospital in New York City, was devastated as he saw the disease attack his tiny charges, and he was determined to find a way to treat it. Early in 1950, when the premature daughter of a colleague developed early retinopathy, Silverman felt that he must do something. Acting "more in desperation than conviction," he began treatment with ACTH, a steroid, which had never been used for retinopathy of prematurity. The results were gratifying: The baby thrived, the early signs of disease receded, and she went home from the hospital with nearly normal eyes.

One successful case was hardly proof, particularly since retinopathy sometimes improves on its own. As Belgian pediatrician Karel Allegaert has written, it is vital that powerful drugs (steroids are a classic example) be tested in children before they are widely prescribed for any specific condition. Still, this was a treatment that seemed to work.

While Silverman and his colleagues at Babies began using steroids routinely, the doctors at nearby Lincoln Hospital were reluctant to use this unproven and potentially dangerous drug. During a period when the Lincoln doctors cared for a total of seven infants with early retinopathy, six went home blind. During the same period, Silverman's team at Babies treated 31 affected infants with steroids, of whom 25 went home with normal eyes.

This was very encouraging, but it still fell short of the proof that would turn steroids from a promising possibility into a validated treatment. The two hospitals had different staffs, different routines, and drew on different patient populations, so it was quite possible that something else was responsible for the better results at Babies. Real proof could be provided only by a "comparative effectiveness trial" that pitted the two treatments against each other, head to head.

To conduct this experiment, the true believers at Babies and the skeptics at Lincoln joined forces and agreed that steroids were to be used for some, but not all, of the babies at each hospital. For months, every time an infant at either Babies or Lincoln developed early retinopathy, the doctor reached into a box containing an equal number of blue and white marbles. The babies now received steroids based not on the doctors' intuition but on the verdict of the marbles.

The results were crushing. Thirty-three infants had completed treatment with steroids—or nontreatment—during the trial. About 20 percent of the untreated babies had gone on to become blind. In the babies treated with steroids, the rate was about 33 percent, and there were more deaths in that group as well. This was not the result Silverman had hoped for, but it put steroids back on the shelf and spared future infants from harm. Other doctors from other institutions tried other treatments. The best of these made no difference; the rest increased the chance of injury or death.

One hope remained: to identify and eliminate the cause of early retinopathy. In this search, dozens of influences on mother and child came under suspicion. Perhaps the mothers had taken a toxic medication during pregnancy, or been vaccinated against smallpox, or been exposed to X-rays. Perhaps the infants had been exposed to too much light, or were deficient in vitamin A, or had been given too many blood transfusions.

Or the culprit might be oxygen. We think of oxygen as the stuff of life, which is true, but it is also a powerful chemical that can be very toxic. Nonetheless, it has always been central in the care of the premature. Its value first becomes obvious in the delivery room, where oxygen in the hands of a pediatrician magically turns a baby from blue to pink. Because oxygen was used increasingly widely in the 1940s, it was natural for pediatricians to wonder if the gas might somehow play a role in this new eye disease. Perhaps the babies with retinopathy had been given too much, or too little, or it had been withdrawn too rapidly.

There were reasons to doubt each of these possibilities, however, especially the idea that the problem was too much. Extra oxygen was obviously essential to these babies' survival, and there was the Tulane experience to consider. The Tulane School of Medicine in New Orleans ran a highly organized program that put as much oxygen into its incubators as any other medical center, yet retinopathy was uncommon there.

Arnall Patz, an ophthalmology resident at Gallinger Hospital in Washington, D.C., still thought that too much oxygen might be at fault, and in 1949 he applied for an NIH grant to study the issue. Routine practice at Gallinger, as at other institutions, was to provide all the premature infants with high oxygen concentrations. Patz proposed a controlled trial in which half of the infants would be maintained at lower concentrations instead.

There were no IRBs in 1949, and research in academic nurseries routinely proceeded without parental consent, but this does not mean that the babies were undefended. NIH reviewers were less than impressed by Patz's underlying theory, and one wrote that, by providing less oxygen, he would "kill a lot of babies . . . to test a wild idea." The reviewers were mollified when Patz promised to provide enough oxygen to keep every baby pink. He was awarded funding in the amount of $4,000, and he soon set about dividing the babies into high- and low-oxygen groups.

The experiment had passed ethical muster at the NIH, but the Gallinger nurses had not been consulted, and many, convinced that the nursery's liberal oxygen policy was saving lives, were horrified. The nursery director was a co-investigator, and while he was on the unit he made sure that the infants received the amount of oxygen they had been assigned. But once he left for the day, some of the nurses went to the incubators in the low-oxygen group and dialed the gas up. They would leave it there all night, then dial it back down before going off shift in the morning.

The nurses' attempt at rescue reduced the difference in oxygen provided, which reduced the likelihood that useful results would be found. The difference in outcome of the two groups, which totaled sixty-five infants, was nonetheless remarkable. In the lower-oxygen group, there were no cases of severe retinopathy, while in the higher-oxygen group, there were seven. Oxygen was blinding babies. This discovery did not solve the problem of retinopathy, because many premature babies will die without extra oxygen. It did make it clear that oxygen could be harmful—critical information that continues to shape neonatal care today.

With this knowledge in hand, investigators were able to answer Silverman's riddle about the increased incidence of disease in the most sophisticated hospitals. Those institutions tended to have the newest incubators and the most vigilant staff, and they did better at giving consistently high concentrations of oxygen. The lower-status hospitals were giving less oxygen, and they were sending fewer babies home blind.

There was one more piece of the puzzle—the Tulane anomaly—and it snapped into place perfectly. Harvard and Tulane were both forcing a lot of oxygen into their incubators, but there was much more retinopathy at Harvard. Careful inspection of the incubators at Tulane showed that they were far from airtight, so oxygen was leaking out nearly as fast as it was pumped in.

THE MOVEMENT TO PROVIDE life-saving treatments to premature babies expanded in the 1950s, and, in 1960, Yale-New Haven Hospital built the first modern neonatal intensive care unit. Others soon followed. But the NICU, with its improvements in every aspect of care for the neonate, was still unable to save many premature infants. The public learned of the limits of care after the only modern birth of a premature baby to a sitting president. Jackie Kennedy delivered Patrick Bouvier Kennedy on August 7, 1963. He was born at 34 weeks' gestation, and although he weighed a reasonable 4 pounds 10 ounces, his immature lungs were unable to use the oxygen he was given. Despite the best care available, he died two days later.

At the time, Jon Tyson was starting his senior year in college with a double major in physics and psychology. Tyson had an analytic turn of mind—a good quality for a young man who wanted to be a pediatrician—and today he is a neonatologist at the University of Texas medical school in Houston. I once asked him how the outlook for similar babies has changed in the last half century.

"The Kennedy baby?" Tyson leaned back in his chair and tented his fingers. "The Kennedy baby would be home in a week."

Better treatments for immature lungs were among the advances that pushed the age of viability earlier, and by 1975 it was possible, albeit rarely, to save infants born at 28 weeks' gestation. Improvements in care continued, step by step. Today, almost all 28-week infants survive, and babies born at 22 weeks and weighing just one pound—small enough to fit in a large soup bowl—can sometimes be saved.

I say "sometimes" because many babies born this early cannot be resuscitated. Further, rescue may come at a cost. Neonatologist Mike O'Shea is following how often major disabilities such as blindness or deafness, severe cerebral palsy, or severe cognitive impairment, defined as an IQ under 55, develop in children who are born extremely prematurely. Of these children, about one-third have no major disability, one-third have one major disability, and the remaining third have two or more. The exact numbers vary, depending on the degree of prematurity, sex (premature females are more robust than premature males), and other factors, and they are improving over time.

When designing his oxygen study, Arnall Patz did not have to contend with an IRB, but today, research to improve survival and reduce disability for premature infants depends on OHRP rulings about parental consent. The conventional model of consent for research with children, in which a parent learns about an experiment and decides whether or not to enroll the child, fits well with some neonatal research, but there are also circumstances in which it is difficult or impossible.

When a woman delivers a premature infant, a pediatrician is normally standing by in the delivery room. The pediatrician works at a table fitted out with equipment that includes a radiant warmer, soft towels, and the equipment for delivering a measured amount of oxygen. For years, the proper concentration of that oxygen was a matter of guesswork. As of 2005, the American Academy of Pediatrics recommended that newborns initially be given 100 percent oxygen, but this recommendation was not based on reliable evidence, and expert opinion was divided. Some neonatologists were convinced that, for these few minutes, it was impossible to give too much oxygen to a baby who was struggling to breathe. Other neonatologists began with room air and added oxygen only as needed, in the hope that lower concentrations would cause less retinopathy without increasing mortality.

Because of this expert disagreement, the amount of oxygen given to premature babies depended on which doctor was on duty; treatment varied from one hospital to another and even from one shift to the next. As in Silverman's steroid study, this was the ideal situation for a comparative effectiveness trial, with higher or lower oxygen levels assigned by a system akin to the Babies Hospital box of marbles. For any individual baby, the risks and benefits of random assignment to an experimental group are essentially equivalent to receiving a treatment based on the equally random habits of the doctor or hospital. Such a study thus generates invaluable data without exposing infants to additional risks. It also raises complex questions about consent.

In 2010, when Vishal Kapadia and his colleagues at Parkland Hospital in Dallas planned a randomized comparison of pure oxygen versus room air as the initial gas used during resuscitation, they saw obtaining consent during labor as a dubious proposition. Some labors are languorous in their early stages, with both prospective parents available and happy to have a conversation with a scientist about an important experiment. Other labors—particularly those that are very preterm—are hectic, with the mother in no mood to converse. With so many women unable to participate, consent before delivery was not feasible. Consent after delivery was impossible, since the experiment lasted only ten minutes and would be completed before the woman had recovered enough to discuss its risks and benefits.

With consent so problematic, another option was not to do the study at all, but this would leave unanswered a treatment dilemma that affected millions of babies every year. Kapadia decided that the best approach was to inform the parents about the experiment after the mother had recovered. Although Kapadia describes this notification as "informed consent," it is not—consent is meaningless when refusal is no longer an option. It is still worthwhile to notify the parents, since it respects their right to be informed about their baby's care.

The regulations recognize that, because of the great variety of circumstances in which research is conducted, it will sometimes be difficult or impossible to obtain traditional consent. IRBs are therefore permitted to waive consent if the research involves no more than minimal risk, the waiver would not harm the subjects, and the research could not practicably be carried out without the waiver. Most IRBs use this waiver authority cautiously, but Kapadia was lucky. His IRB ruled that it was not practicable to do the study without a waiver of consent, and it permitted the research to proceed so long as the parents were notified of the experiment after delivery.

Kapadia moved forward, and in just six months he had impressive results. Babies who were started on room air spent less time on ventilators and were less likely to develop lung disease. They were also less likely to die or go blind, although these results did not reach statistical significance. This study ended the era of guesswork: Babies struggling to take their first breaths are more likely to live and thrive on room air. Even a brief exposure to pure oxygen is toxic.

THERE REMAINED THE OPEN QUESTION of what to do after the infant was moved from the delivery room to the NICU. Patz had proved that too much oxygen blinded babies, but too little was also damaging. In time the tradeoffs became clear: Too little oxygen can cause brain damage or death, while too much can cause both retinopathy and chronic lung disease. The ideal level, one that would balance the risk of these complications, was unknown.

New technology made it possible to address these uncertainties. Pulse oximeters, which became available in the 1980s, use sensors taped to an infant's palm or foot to measure saturation—the level of oxygen the blood is actually carrying—up to a maximum of 100 percent. It is good to know how much oxygen is in the incubator, but even better to know how much is in the baby. This could now be accurately determined. Of course, nobody knew precisely what level was ideal—that could become known only with a trial using the new oximeters.

There had been a variety of studies of oxygen in premature babies since Patz's time, but none of their results could be extrapolated with confidence to the present. William Silverman, the pediatrician who had shown that steroids were harmful for premature babies in the 1950s, was still active half a century later, and he summarized the state of knowledge as of 2004: "there has never been a shred of convincing evidence" to guide clinicians as they administer oxygen in the NICU, since the optimal level "was, and remains to this day, unknown." In this evidentiary vacuum, the American Academy of Pediatrics recommended oxygen saturation levels of 85 percent to 95 percent, but this was more a matter of intuition than evidence. One end of this range might be better than another. The new oximeters allowed neonatologists to plan an experiment to find out.

To reach statistical significance, such a study would need to be massive, with more than a thousand babies. No medical center had enough resources, or enough babies, to bring it to completion. Fortunately, the Neonatal Research Network had been created to tackle major problems like this. The Network was established in 1986 by the NIH's Eunice Kennedy National Institute of Child Health and Human Development. It comprises NIH coordinating staff, a statistical center, and fifteen academic medical centers across the country, some of which have more than one hospital. Its scientists include many of the country's most distinguished neonatologists (Jon Tyson is a member). It has deep expertise in evaluating "nonvalidated therapies"—treatments that, although they are in common use, have never been rigorously studied.

As it contemplated launching an oxygen trial, the Network's first step was to form a study group to conduct an exhaustive search of the published literature. It found no trend, no pattern, only a jumble of experiments, often decades old, that provided no reliable guidance for clinical care. It noted that neonatologists had "strong beliefs regarding oxygen management" but that these beliefs often pointed in opposite directions.

The Network established a committee to draw up a plan for a trial to be conducted with premature infants born between 24 and 27 weeks' gestation who weighed as little as one pound. Its preliminary proposal was critiqued by experts in physiology, research methods, and statistics. Its methods were refined, and a second component was added to compare two different ways of helping newborns with the work of breathing. The final plan, which was dubbed SUPPORT, compared infants given oxygen at the high and low ends of the recommended range (91-95 percent versus 85-89 percent). They would be maintained at those levels for as long as they needed extra oxygen.

The SUPPORT planners remembered Patz's troubles with the nurses on the night shift. It therefore called for "education," which meant persuading nurses that their institution's customary oxygen levels, which they had zealously maintained for years, were based on opinion and habit, not on evidence. If their NICU used a higher level and another NICU used a lower one, it was ethical for both NICUs to try both levels to see if one was better.

Once the study protocol was finalized, the next task was to obtain IRB approval at each institution. When Kapadia's IRB reviewed his study of the use of oxygen in the delivery room, it had approved waiving consent and instead informing the mothers after they had recovered from the delivery. If the IRBs for SUPPORT followed a similar approach, consent would need to be waived only for the first day or two. At that point, there would be an opportunity to approach the parents and ask for consent for their baby to continue in the trial. Infants whose parents opted out would be given whatever level of oxygen was the local custom for the remaining three or four months of their hospitalization.

The SUPPORT investigators, like Kapadia, rejected the idea of trying to obtain consent from women in active labor. There was, however, a more cumbersome way to obtain consent before the baby was delivered. Pregnant women in threatened preterm labor are often hospitalized. They are put on bed rest while their contractions sputter unpredictably and doctors try to coax the pregnancy along for a few more precious days or weeks. While it is an anxious time for the mothers, it can also be quite tedious, leaving ample opportunity for them to talk with the scientists. The investigators could explain the study to these hospitalized women and, for those who consented, plan to enroll their babies in SUPPORT after they delivered.

This reasonable-sounding plan had a serious drawback: Most women admitted with preterm labor succeed in carrying their pregnancy past the period of greatest risk, which would make their babies ineligible for SUPPORT. That is good news for the babies but a bad return on the time spent obtaining consent. Consent before labor was therefore possible, but not practicable, a word that should trigger the waiver of consent that Kapadia used. However, all of the IRBs—and there were twenty-three involved—rejected this option. In a decision that OHRP approved, they all required that consent be obtained before the woman went into labor. The Neonatal Research Network would soon learn the extravagant cost of this requirement.

SUPPORT began enrollment in 2004. In order to obtain prior consent, specially trained nurses screened thousands of women who were at risk of delivering early. Four out of five did not deliver an eligible infant, which wasted both their and their nurses' time and slowed completion of the study.

Four years later, the results were in. The data showed that the higher-oxygen group had more blindness but also better survival, at 19.9 percent versus 16.2 percent, a finding that was statistically significant.

This was valuable new knowledge. But the consent requirements had drained $200,000 from the SUPPORT budget and delayed the study's completion by two years, depriving neonatologists of knowledge that the higher oxygen level was a better choice for the 200,000 very low birthweight infants born during that delay.

The cumbersome consent process in SUPPORT imposed one more cost. When the IRB system was created, it was generally assumed that subjects who participated in research were making a sacrifice. By the 1980s, it became obvious that groups may be harmed by being *excluded* from research. When a new treatment is tested only in white men, for instance, the results may not be generalizable to women and minorities. In 1987, NIH instructed investigators to include women and minorities in clinical studies, and in 1993 Congress made inclusive enrollment mandatory for NIH-funded research.

Inclusive enrollment was settled policy at the time of the SUPPORT trial, yet the study's comprehensive consent process undermined equal enrollment. The mothers of infants who were not enrolled were less likely to be white, have medical insurance, have completed high school, and have received prenatal care. Delaying consent until soon after birth would have likely resulted in infant participants who were more representative of extremely premature babies generally. There is no reason to think that disadvantaged mothers were less willing to participate; rather, they were more likely to arrive at the hospital on an emergency basis with too little time to complete the consent process. Requiring consent before birth thus resulted in findings that were tailored to the children of relative privilege.

These heavy costs suggest that the consent process in this case was neither practicable nor in the best interests of parents and premature babies, then and now. And, because it skewed the sample toward more privileged parents, it also violated the IRB system's own aspirations for inclusion and diversity.

FOUR YEARS AFTER SUPPORT WAS COMPLETED, OHRP launched a fresh attack on the study's consent form, IRBs, and scientists. In a February 2013 letter of reprimand addressed to the University of Alabama at Birmingham, the agency asserted that every consent form used in the study was defective. The letter demanded corrective action for this alleged failure and left open the possibility of future sanctions for all concerned.

As we've seen, before SUPPORT was launched, each NICU followed its own routine regarding how much oxygen to provide, and every infant was at risk from the unpredictable complications that can result from too much or too little oxygen. SUPPORT was a comparative effectiveness study, with every infant receiving care that was routine in some NICUs at the time. The UAB form therefore indicated that the risk added by the experiment was minimal. OHRP sharply criticized this approach, writing that the consent form should have warned of the "substantial risks" of participation, including "the reasonably foreseeable risks of blindness, neurological damage, and death."

The IRB members were dumbfounded. These men and women, doing work that was usually unpaid and never glamorous, had taken every precaution they could imagine, including requiring the scientists to obtain consent before the mothers delivered. Now, nine years later, OHRP was accusing them of approving consent forms that were deceptive. Not for the first time, the agency was interpreting IRBs' reasonable decisions as a blatant violation of the regulations.

The scientists were equally stunned. The Network had designed a study to compare treatments that were already in common use, shown that one was superior, and improved the care given to premature infants around the world. This remarkable achievement was the result of years of work by thousands of scientists, doctors, and nurses. OHRP's attack on this completed experiment—an attack the scientists considered unjustified—also put their other, ongoing studies in jeopardy.

A careful study with valuable results can still use a defective consent form, so OHRP was within its rights. And, to give the agency credit, there was a kernel of truth in its allegations, since any change in a NICU's routine oxygen practice might change the outcome for any individual infant. But the agency distorted that kernel in order to justify a misleading campaign, one that was amplified and further distorted in the media, advocacy group press releases, and court filings. The melee that followed bruised all of the participants, including the scientists, the IRBs, and OHRP itself. Nor were the parents spared, since some became falsely persuaded that they had made a terrible mistake in enrolling their children in the study.

Before it could argue that the consent form was defective, OHRP had to overcome a regulatory barrier. The treatment of disease is never free of hazard, so the regulations stipulate that when IRBs evaluate risks, they "should consider only those risks and benefits that may result from the research (as distinguished from risks and benefits of therapies subjects would receive even if not participating in the research)." In order to bypass this requirement, the agency argued that an infant who participated in the study might receive more oxygen than would have been provided in routine care, which could result in blindness. The infant also might receive less oxygen, which could lead to brain injury or death.

To understand OHRP's reasoning, imagine a specific infant—let's call him Baby X—born at 25 weeks and weighing a pound and a half. He would have been better off if his parents had never heard of SUPPORT, since he would have done well on the NICU's customary oxygen level and gone home with no problems. Unfortunately, his parents enrolled him in SUPPORT, and he received more oxygen, developed retinopathy and became blind, or less oxygen, suffering brain damage or death. OHRP's letter takes this infant's injury as proof that the IRBs should have required a stern warning in the consent form.

The letter fails to mention another infant, Baby Y, who might have been born at the same hospital and treated in the same NICU. Had he been given that unit's customary oxygen level, he was fated to suffer blindness, brain damage, or death. Fortunately, his parents enrolled him in SUPPORT, and he received less oxygen, sparing him blindness, or more oxygen, thus avoiding brain damage or death. Enrolling in SUPPORT would thus harm one baby and benefit the other. When it ignores Baby Y and focuses only on Baby X, OHRP gives a distorted picture of the study's risks.

OHRP also omits another crucial consideration. Most parents understand that the benefit or harm from participation in an experiment is unpredictable. Still, they want to know if enrollment is a smart bet, and looking back they will care less about what the consent form said than about whether enrollment actually helped or harmed their baby. Because SUPPORT had been completed at the time of OHRP's action, the answer was already known: Babies in the study generally did at least as well as they would have with routine care. OHRP chose not to mention this vital fact.

By hammering on the risks that it claimed should have been disclosed, the agency wrapped the study in an air of danger without ever claiming that it was actually riskier than ordinary care. This sleight of hand, with its stealthy decoupling of exaggerated risk from any actual injury, led to confusion among journalists and a raft of troubles for the scientists. Perhaps OHRP intended to buttress its reputation as an aggressive defender of subjects' rights.

Word of OHRP's action spread rapidly, with the *New York Times* and other media providing updates on the controversy. The *Times* editorial board praised the agency and condemned the medical centers involved for a "startling and deplorable" failure to disclose the study's dangers, including the risk of death. A few years earlier, the agency's attack on Berenholtz and Pronovost's intensive care unit study had been pilloried in the same newspaper. Its praise of OHRP's attack on the infant oxygen study must have been gratifying reading at agency headquarters.

SUPPORT was conducted at hospitals across the country, so local media had an easy local hook. Perhaps because UAB was the lead institution, some of the other hospitals apparently did not realize they would be spotlighted for their participation, and it was easy for reporters to wonder if they were hiding something. A reporter for the *Dallas Morning News*, writing on deadline, called the University of Texas Southwestern, requested an interview with one of the SUPPORT investigators, and was turned down. The school did promise a written explanation of what happened in the study, then failed to deliver it. The next morning, the headline was, "Dallas researchers fail to explain why parents in national baby study weren't notified of fatal risks." This prompted the school to produce a statement denying that any babies had been harmed. The paper ran another article, its headline suggesting skepticism of this claim: "UTSW enrolled 73 infants in controversial study; 17 of them died."

In Dallas and elsewhere, the scientists struggled to explain the study in terms the public could understand. Their difficulty was not in the facts themselves. A death rate of 17 out of 73 is just what would be expected among these infants, who were kept alive by the thinnest of margins whether they were enrolled in research or not. For years, the neonatologists had been giving more or less oxygen based on the slimmest of hunches and without permission from anybody. To conduct SUPPORT, they had simply agreed to make oxygen administration systematic and keep track of the results. They discovered that more oxygen led to increased survival, a result that none of them had predicted—which is precisely why the trial had to be done. If they had known the answer, they would have studied something else.

This rational explanation was not always effective against the indignation that OHRP had aroused, particularly for onlookers who assumed that babies were safe in routine clinical care but exposed to hazard if they were enrolled in an experiment. Without a direct rebuttal of the agency's charges, the scientists' denials seemed merely defensive. They needed to reverse engineer OHRP's letter of reprimand and reveal the agency's sleight of hand. Pediatrician John Lantos led the effort to make their case more effectively.

Advocacy groups soon joined the fight on OHRP's side. Public Citizen, the organization founded by Ralph Nader, expressed horror that the infants were deprived of oxygen tailored to their individual needs. It was confident that most parents would have rejected the study had they known of its risks. The private, nonprofit group Alliance for Human Research Protection wrote of babies who had been deliberately sacrificed.

In the next and probably inevitable step, the publicity attracted a team of lawyers, who filed a lawsuit on behalf of hundreds of infants who had been enrolled in the trial. The suit charged the UAB investigators and individual members of its IRB with negligence, failure to obtain informed consent, and wrongful death.

The scientists were now under attack from every point of the compass. Even neonatologists who were not involved in the study put other research on hold as they waited to see what other projects— planned, underway, or completed—OHRP might target next. Nor was the fear limited to neonatology, as OHRP's enforcement action raised broad questions about research in every field of medicine.

In June, the NIH top leadership reached out to help. Director Francis Collins co-authored an editorial that sharply criticized OHRP and observed that the agency's actions had "alarmed some of the parents of infants who were in the study, confused the biomedical research community, and befuddled IRBs." The agency responded that it wished "to emphasize that OHRP does not and has never questioned whether the design of the SUPPORT study was ethical" and was only concerned about whether the parents had been properly informed.

OHRP put its compliance action against UAB on hold, but did not dismiss it, nor did it let the matter rest. In August 2013, it convened a meeting to discuss SUPPORT at the Department of Health and Human Services. OHRP's head, Jerry Menikoff, and two other officials sat at a table at the front of the room. During the course of a day, the panel heard from twenty-seven people, including ethicists, scientists, and parents. Public Citizen took advantage of the meeting to hold a press conference during the lunch break.

Some of the ethicists defended the study; others considered it an abomination. George Annas, who spoke at the meeting, would later write that one of the problems was "denial of death, illustrated by the inability of researchers, IRBs, and supporters of SUPPORT, to acknowledge the fact that the study itself could put the newborn subjects at increased risk of death."

The scientists were eager to respond. Lantos explained that the choice between routine care and participation in a study always involves trade-offs, so consent forms should help people "understand that by consenting to research, they may be avoiding some risks and exposing themselves to others." This balanced analysis is just what OHRP rejected when it focused on Baby X and ignored Baby Y.

Lantos also walked the audience through problems with therapies, like oxygen in the NICU, that are used every day but have never been properly validated. When experiments compare two treatments that are already in routine use, consent forms that emphasize experimental dangers "are not empowering people to make informed choices, they are scaring them into making uninformed ones. ... When consent forms overstate risks of research, make no mention of the risk of conventional therapy, [and] don't say research subjects might be better off than patients who are not in studies, they are misleading and dangerous."

These arguments may have been persuasive to scientists in the audience, but they made no discernible difference to the parents, who all appeared to believe they had been deceived and their babies had been harmed. Shawn Pratt, who was there with his wife Carrie, said that in their minds the study turned Dagen, their daughter, into "a subject of an experiment."

Tyson, who had been one of the study investigators, said that in his opinion, every premature baby born while the trial was being conducted was a research subject. Babies like Dagen who participated in SUPPORT were, by definition, in an experiment. But, Tyson went on, so were the babies who were not enrolled, since they were participants in the uncontrolled, unmonitored experiment that is routine clinical care when the best treatment is unknown. The main difference is that in routine care the amount of oxygen given is determined by the customs of individual doctors or hospitals, just as the use of steroids in New York in 1950 depended on whether the premature infants were at Lincoln or Babies.

In routine care, nobody tries to correlate the amount of oxygen given with the complications that occur, because there are too many other possible explanations. The SUPPORT investigators broke new ground by keeping everything else the same and assigning higher or lower oxygen levels by the modern equivalent of Silverman's box of marbles. Tyson also pointed to some encouraging data about the babies in the study: "there is no evidence that [they] had a worse risk adjusted outcome than other similar babies not enrolled."

In its letter of reprimand, OHRP had stressed the increased mortality of babies receiving less oxygen, and one detractor after another echoed the point at the meeting. Tyson's observation about risk-adjusted outcomes should have put this concern to bed, but the study's critics either disregarded or did not understand his comment. To be fair, the analysis requires some unpacking, but it is worth the effort. Here is what he meant and why it matters.

Hospitals in the Neonatal Research Network keep meticulous data about all premature babies in their care, whether or not they are enrolled in any particular study. During the trial, 4369 infants treated in the Network's hospitals were eligible for SUPPORT. They fell into three groups: those who were enrolled in the study and assigned to high oxygen, those who were enrolled and assigned to low oxygen, and those who were not enrolled. By every important measure, those who were enrolled, as a group, had better outcomes than those who were not enrolled. Some of the differences did not reach statistical significance, and the two groups were not the same—the babies who were enrolled had, as a group, some initial advantages over those who were not—but the raw data are still striking. Babies enrolled in the study, including both high- and low-oxygen groups, had less lung disease, less retinopathy, less dangerous intestinal infections, less cerebral hemorrhage, and less brain injury. Less, in other words, of every devastating complication short of death. If you were looking for safety, the study was the place to be.

The mortality data were even more powerful. It is true, as OHRP repeatedly pointed out, that more babies in the low-oxygen group died, and that is the agency's justification for emphasizing that parents should have been warned about the risk of death. Vera Sharav, of the Alliance for Human Research Protection, stressed this point at the meeting. Noting that 237 of the infants in SUPPORT had died, she said, "not a single IRB-approved consent form mentioned death. . . . How many deaths," she asked, "can be attributed to SUPPORT?"

We can't unspool time, cancel the study, and see how the infants would have fared if they had not been enrolled. But thanks to the Network hospitals' detailed records, we can take the 1,316 babies who were enrolled and compare them with the 3,053 other babies, treated by the same doctors and nurses in the same NICUs at the same time, to get an idea of the study's likely impact. Mortality in the three groups—enrolled and given high oxygen, enrolled and given low oxygen, and unenrolled—was 16.2 percent, 19.9 percent, and 24.1 percent. If we proportionally reduce the 3,053 babies who were not enrolled to match the size of the group in the study, a mortality rate of 24.1 percent amounts to 320 deaths. Among babies who were enrolled, there were 107 deaths in the high-oxygen group and 130 in the low-oxygen group, for a total of 237. At this first level of approximation, it would seem that the answer to Sharav's question is that enrolling in SUPPORT saved the lives of 83 babies and that the consent form should have emphasized life, not death.

I call these numbers an approximation because it's impossible to know if the better survival of babies in the study was due to their participation. Going into the trial, they had some advantages over those who were not enrolled: On average, their mothers had received more prenatal care, antibiotics, and medication to speed lung maturation. An analysis that adjusts for these advantages shows no statistically significant benefit from enrollment in SUPPORT. Tyson was making this point more succinctly when he said that "there is no evidence that [they] had a worse risk adjusted outcome." In other words, we don't know that participation in the trial helped, but there is no evidence to suggest that it caused harm of any kind. All the heated rhetoric about the danger of death in SUPPORT was baseless.

Around the same time, three large studies abroad were also looking at the ideal level of oxygen, and they all found the same pattern. Despite the general finding of greater mortality in the low-oxygen group compared to those given higher oxygen, in none of these trials did babies enrolled in the low-oxygen group have higher mortality than babies who were not in the study. As Lantos pointed out, findings like these are not surprising, since being in a clinical trial can benefit participants in both arms of the study, whether they get treatment A or treatment B. The reasons are not fully understood, but in the case of SUPPORT, close attention to oxygen levels may have reduced the wider swings that can occur in routine care.

These data, and the analyses by Lantos, Tyson, and others, provided the scientists with welcome support. The NIH had already persuaded OHRP to take a step back, and time brought more welcome news when an Alabama judge granted summary judgment in favor of UAB and dismissed the lawsuit. Lantos, reporting the ruling in the *New England Journal of Medicine*, wrote that the judge had found that "no reasonable person could conclude that being in the study caused harm." The triumphant headline read, "Vindication for SUPPORT."

"Vindication" is an overstatement. Ethicists continued to attack the study, the money UAB had spent to defend itself in court was gone, the *New York Times* did not retract its editorial, Public Citizen continued its attack, the parents remained convinced their babies had been injured, and fear and confusion among scientists over who might be targeted next remained. OHRP itself yielded no further ground. Asked to comment on the decision, Menikoff said tartly, "The consent form was inadequate at the time of the study, and the court ruling doesn't change that."

MOST OF THE LITERATURE about the SUPPORT debate focuses on the consent form and the effects of more or less oxygen. I have never found an article that discusses the debate from the point of view of the parents and their children, yet their experience—the experience of more than a thousand families—is the controversy's moral core.

The parents and the professionals were in very different situations. The scientists, ethicists, and lawyers argued over issues in their professional areas of expertise; at the end of the day, they could close the office door and go home. Few had spent months by their child's incubator in the NICU. The parents, in contrast, knew little or nothing about ethics or neonatology. They simply wanted a healthy child, and if their decision to enroll their baby in SUPPORT had been a tragic mistake, they would blame themselves for the rest of their lives. The impact of the debate on the families is therefore worth thoughtful consideration.

To understand their reaction to the SUPPORT debate, we need to understand what life was like during the fraught months when their baby was in the NICU. So far, we have focused on oxygen in the care of premature infants, but many other factors are critical as well. Keeping an extremely premature baby alive and giving that baby a chance to grow requires meticulous attention to the infant's temperature, blood pressure, intravenous fluids, nutrition, lighting (too much isn't good), movement (ditto), medication (perhaps vitamins, iron, antibiotics, caffeine), and much more.

Doctors write the orders for all this. Nurses do the actual work, checking vital signs, providing feedings, starting intravenous lines, administering medication, and coordinating these activities so that the babies have enough time to sleep undisturbed. Nurses also respond to the NICU's constant alarms by fixing plugged tubes, adjusting oxygen levels, and tapping the feet of infants who have temporarily forgotten to breathe. This is a high-tech environment crammed with incubators, respirators, and monitors.

The NICU is also a human community of people who are doing everything in their power to sustain a frail and precious thread of life. The parents play an important role, and nurses coach them in how to feed and soothe a two-pound baby without accidentally pulling out a wire or tube. Family involvement makes a difference, since babies whose parents are engaged in their care do better and are ready to go home sooner. There are articles that lay out the scientific reasons why these infants benefit from parental care, but I see no reason to look beyond the universal need to love and be loved.

This work, both high and low tech, goes on day after day, with a typical hospitalization of three or four months for an extremely premature baby. A lucky few cruise through their hospital stay and go home as ready as any other baby to meet the world. They are the exception, since a majority of babies who spend months in the NICU go home with some degree of long-lasting health issues. Most families rise to this challenge, finding resources to help with the visual impairment, the cerebral palsy, the lung disease. And they find as much love and joy with their child as other parents—perhaps even more.

When they look back on those months in the NICU, so full of hope and anxiety, most parents are sustained by two convictions. They believe that the choices they made during this stressful time were sound and promoted their baby's welfare. And they believe that the doctors and nurses, whether they were providing standard treatments or doing research to find out how those treatments could be improved, were giving the best care humanly possible.

Any federal official who wanted to challenge a study that had been a part of this intense period for so many families would face a delicate task. SUPPORT had been completed years before, and thousands of parents had never suspected there was anything wrong with it. The children, aged between four and nine when OHRP launched its campaign, had no reason to believe that, as infants, they had been deliberately exposed to harm. Menikoff is not just director of OHRP, he is a doctor, so he knows that when parents have made reasonable choices and their children have gotten good care, it is wrong to imply that the parents or doctors were somehow at fault when their children struggle with disabilities.

Almost all of the SUPPORT parents would be vulnerable to an attack on their choices and the care their infants received. Few babies emerge from their time in the NICU in perfect health, and while survivors' visual problems, cerebral palsy, or lung disease could always be worse, they could also be better. "If we had made a different choice, our child wouldn't be struggling today" is an easy thought to have and a hard one to forget.

Menikoff could have respected their situation in two ways. The simpler would have been to use professional channels to communicate his agency's concern about the consent form to the institutions involved, a quiet path that would be unlikely to mislead and alarm the parents. He chose, instead, to seek the widest possible public audience. If he had made it crystal clear that while he considered the consent form to be defective, enrolling in the study was a choice any loving parent would be happy to make, he would still have met his obligation to the families.

Menikoff did nothing of the kind, and perhaps he felt that he had no obligation to the families. The parents were never research subjects, after all, and the children had not been in the study for years. But it seems to me that if he was so concerned about the rights of the families at the time, he should not have harmed them later. Yet he did all he could to encourage the twin misperceptions that the study was dangerous and that the parents' decisions had harmed their babies. Neither was true.

At the public meeting, Lantos emphasized the guesswork involved in routine care and the potential benefits of study participation, and he asked Menikoff this pointed question: "Is it permissible now to say in the consent form, for a study like SUPPORT, babies in the study might be better off than babies who are not?"

Menikoff answered, "I don't want to give a personal viewpoint as opposed to an official regulatory position."

His evasion was shrewd, since admitting that parents were smart to enroll their babies in SUPPORT would have undercut his entire campaign. Menikoff had never argued that the babies were injured— he left that to the media and Public Citizen. But to openly concede that the babies had generally done well would have been to abandon the agency's most emotionally persuasive tool. The public outrage over the study was fueled far less by any failure to inform the parents than by the injuries supposedly inflicted on the babies.

The parents' testimony, at the meeting and at the press conference organized by Public Citizen, was gripping. Sharissa Cook said that her son's doctors, by not disclosing what they knew about the study's risks, took away her ability to make an informed decision.

"Had I known the full extent of the study, I would not have given my consent. . . . I unknowingly placed my son in harm's way. . . . I trusted them with my baby's life." Shawn Pratt said that he and Carrie had agreed to the SUPPORT study only after a great deal of thought. They had thought that was a reasonable decision. And then, years later, "new information" had come in. "Imagine our surprise," he said, "as we learned about the risks associated with the SUPPORT study." He felt cheated: "We were not told that there were predictable increases in risks to our baby."

The press conference lineup included a lawyer, an ethicist, the Public Citizen sponsors, and Shawn Pratt, all men wearing dark suits, but Public Citizen's most effective witness did not say a word. The center of attraction, the person on whom all eyes were turned, was six years old, four feet tall, and wore pigtails and a colorful sundress. Dagen's thin limbs, and the braces on her legs, seemed to prove what OHRP could only imply: that SUPPORT had injured her. Carrie spoke of her distress over what she believed was a deceptive consent form and her remorse over enrolling her daughter in the study. Tearing up, she struggled to finish. "We're very blessed that Dagen survived and has her eyesight, but every day that she cries because she's different, it just kills us." She had two questions: "Why were we not totally informed of the risk? And why is omitting information not considered lying?" Public Citizen's article about the case is even headlined, "Outrage of the Month."

The real outrage here is how OHRP launched a deceptive campaign and Public Citizen amplified it, despite the pain they had to know it would cause the families. The families suffered not because of the experiment, but because of the investigation. The story of these parents and their children has not previously been told. It is an indictment of a government agency that harmed the people it is charged to protect.

Agency actions like this harm us all, of course. We all suffer when officials wielding federal power kneecap investigators who are struggling to expand medicine's limited evidence base.

OHRP'S ASSAULT ON THE SUPPORT STUDY is not an isolated event. As we have seen, under Gary Ellis's direction, OPRR imposed heavy sanctions on more than a dozen research universities for faults as minor as uncertainty about whether a quorum was present at every moment of every meeting of the IRB. Under Ivor Pritchard, OHRP threatened, without a whisper of justification, to shut down research in intensive care units that was saving thousands of lives and millions of dollars. In cases that are less notorious but just as important, the agency has repeatedly punished IRBs for decisions that are entirely reasonable.

This pattern cannot be changed under the current system, because OHRP's incentives, which are baked into federal law, all favor aggressive oversight. Fortunately, the law can be changed and the system reformed.

# 10

# Failure

H ENRY BEECHER WAS RIGHT TO PRESS FOR REFORM, for scientists will always need oversight when they put human subjects at risk, and James Shannon was right to implement such a system.

Shannon succeeded. I know of just five subjects who have unexpectedly died as a result of their participation in American research since 1999. Given the millions of subjects who have enrolled in studies in that period, one conclusion is clear: the IRB system helps make medical research very safe.

Unfortunately, the system's failure is equally clear: We pay a staggering cost for this safety.

Before I go further, I want to be clear that oversight always comes at a cost, in terms of dollars spent and research delayed. We should be happy to pay this price so long as the system is run efficiently. The current system, however, is so inefficient that its costs cannot be justified.

The most obvious cost is measured in dollars. NIH research grants typically allocate about 50 percent or more of funds to cover institutional overhead costs, which cover utilities, building maintenance, and administrative and support services. Grant recipients do not make public how they distribute these funds, so how much of this is devoted to IRB operations is unknown. We do know that in the 2000s, each medical school (there are 141 in the United States) spent about $1 million on its IRB. Other institutions—hospitals, universities, research centers, military units, and others—support their own committees.

I think of these costs as an ethics tithe, but whether the total is more or less than 10 percent, the tally, summed across the many thousands of investigations conducted each year, is high. In 2020, the NIH spent over $40 billion on medical research. If half of that research involves humans, and if five percent of that half goes to IRB costs, that is about $1 billion a year; if it is ten percent, it is $2 billion. Whether the actual amount is more or less than a billion dollars, that is money that was intended to fight premature birth or breast cancer but is spent on IRB photocopiers, administrative salaries, proposal review software, and the time of scientists as they slog through the system.

The more significant cost, however, is the lost or diminished lives that result when breakthroughs are delayed or prevented. And, because of the anonymity of the people who suffer when science is damaged or delayed, it is a cost that goes largely unrecognized. Carl Schneider and I, building on Rory Collins's observations, estimated that about 9,000 people died because of the delay in the ISIS-2 heart attack study. Those individuals cannot be identified. We cannot publish their photographs and tell the story of their shortened lives. Yet they vastly outnumber the handful of individuals who have suffered harm as a result of participating in research, and whose story has been repeatedly told. As sociologist Carol Heimer writes, "Because research on human subjects is how we produce the innovations that improve health, reduce morbidity or mortality, and alleviate human suffering, preventing or delaying research results in vastly more suffering and death than occurs from researchers' ethical lapses."

The comprehensive consent process in the SUPPORT study of oxygen for premature infants exacted high costs in time, money, and in the enrollment of babies who were not representative of premature infants more generally. The human costs continued with OHRP's deceptive attack on the SUPPORT study. This was a public relations win for OHRP, which showed the world how tough it is. But it gave thousands of parents reason to believe, falsely, that they had enrolled their child in an experiment that caused the child's death or disability.

You might wonder if I am exaggerating the IRB system's negative effects, since medical advances continue to make headlines and save lives. With those billions poured into the field, of course research will make new discoveries, even after the IRB system has taken its toll in money and time.

But what scientists tell me is that the system channels research into areas and methods that they expect their IRB to approve. They often add that their greatest regret is about work they never even proposed because IRB approval was doubtful. I have not included those stories, for I did not want to base this book on hypotheticals. It would be unfair to the IRB, which after all might have been willing to approve the experiment, and to the reader, who would be asked to take it on faith that the study, once completed, would have been valuable. Yet even IRB-approved research is delayed and damaged.

The development of Covid-19 vaccines was conducted, as the program leaders boasted, at Warp Speed. Less newsworthy research for common killers still proceeds at horse-and-buggy speed. Heart attacks alone claim 655,000 Americans every year, with an increase in the proportion occurring among people aged 35 to 54, especially women. Our drugs for Alzheimer's disease can, at their best, only slow its progression. Our drugs for schizophrenia are for the most part tranquilizers that offer limited benefit. Cancer of the lung, pancreas, and liver are as remorseless as ever.

This is what I have in mind when I say that the IRB system, for all its talk of ethics, misses the moral value of medicine's struggle against suffering and death. An ethics that praises a passerby for rescuing a drowning toddler should also smile when that person contributes to our general welfare.

Make no mistake, experimental subjects still need protection, so we still always need a system to oversee medical research. Just not *this* system. Until we reform it, we will pay the price of failure in the coin of suffering.

Regulators (like the officials at OHRP) and most ethicists disagree. They believe that the current system provides a proven solution to an important problem, and that we tinker with that solution at our peril. Because federal officials control changes to the regulations, and ethicists dominate the public discussion about IRB operations, their joint opposition is a formidable obstacle to fundamental reform.

Let's start with the ethicists. In the last half-century, they have created the academic discipline of research ethics, with endowed chairs, specialized journals, and experts who comment to the media and legislators. Ethicists often make their name by publishing articles that identify modest changes that might improve the system, while generally assuming that something very like the present system is the only way to manage the risks of research. The notion of fundamental change is anathema.

Will van den Hoonaard, of the University of New Brunswick, is a founding member of Canada's Panel on Research Ethics. He has been to IRB conferences in both Canada (where the process is called ethics review, conducted by ethics committees) and the United States. He found these conferences to be quite unlike most academic meetings. On both sides of the border, "there were no critical challenges posed about the ethics review system itself. No critical self-reflection about the work and nature of ethics committees. Participants of both conferences lived in a bubble of compliance. Unlike academic conferences, there were no voices of dissent, nor were any expected."

Nor are scientists expected to dissent. Official education is part of the effort to ensure their submission. Taxpayers do not have to pass an annual exam on the importance of the Internal Revenue Service, nor are motorists quizzed on why it is wrong to drive drunk. But IRB members, and the scientists they regulate, must annually take a course and pass an exam (usually online) in research ethics. This course usually covers some topics that anyone would recognize as relevant to ethics, such as the definition of autonomy or of beneficence. There is some material that is pure regulation, such as the legal requirements for informed consent.

But some of the material has the flavor of indoctrination. It reviews the abusive experimentation, particularly the Tuskegee syphilis experiment, that led to the development of the IRB system. It emphasizes how the IRB system pursues ethical goals, including preventing harm to subjects and promoting justice. And it portrays the IRB as ethics in action. This all implies, not very subtly, that an investigator who challenges an IRB ruling is at best ethically challenged, and at worst morally defective and perhaps racist to boot.

Sociologist Howard Becker provides this summary: "What began years ago as a sort of safeguard against doctors injecting cancer cells into research patients without first asking them if that was OK has turned into a serious, ambitious bureaucracy with interests to protect, a mission to promote, and a self-righteous and self-protective ideology to explain why it's all necessary."

Research is the key to a successful career in universities and medical schools, and investigators are vulnerable to reprisals should they speak of the system's deficiencies. Therefore, while scientists know the harm it does, most are too prudent to complain. Legal scholar Scott Burris writes, "Few are 'against' ethics; no one wants to face the inevitable accusations of moral defect and more or less bald comparison to Nazi doctors that follow criticism of the bioethics project."

Dale Carpenter, of the University of Minnesota Law School, writes, "The *in terrorem* effect of IRB review chills serious critical analysis by the very people—researchers—most familiar with how IRBs operate." Coe is not silenced by this threat, noting acidly, "We

who live in the IRB Gulag openly despise the IRB process ...."
Nonetheless, he "fears retribution, secret retribution meted out and
veiled in procedural details."

And the IRB is ideally positioned to mete out secret retribution,
since the punishment can be presented by the IRB as business as
usual—demanding changes that diminish a study's value or extending
protocol review from a month to three or four months. Scientists in
the IRB doghouse may find themselves spending months responding
to a "routine" audit. Since the system punishes every investigator, if
only by the time spent to fill in forms and the time waiting for
approval, the question is only one of severity.

Retribution for disrespect is justified by the IRB belief system.
Robert Amdur, in his *Institutional Review Board Member
Handbook*, advises IRB members that investigators who criticize the
IRB in the performance of its duty "demonstrate a lack of respect for
the system of protecting research subjects." Such disrespect "is an
important finding that should be taken into account when making
IRB determinations." And Robert Levine, the Yale IRB's long-time
chair, shows how it is done. Yale's IRB "nearly never labels a protocol
disapproved. The way disapproval is accomplished *de facto* is that
requirements for revision are made that the investigator chooses, for
any of a large variety of reasons, not to accept."

The talk of ethics is a boon to federal regulators. Although
Shannon had no use for formal ethics, federal officials charged with
IRB oversight were undoubtedly delighted to have distinguished
philosophers agree that the work of the IRB system embodies noble
ethical principles.

Whether its work is considered ethics or not, OHRP controls the
levers of power. Although I believe that the agency has repeatedly
fumbled its supervisory responsibilities, it has handled pressure to
reform the system with great skill. It is well aware of researchers'
complaints, and in 2011 it indicated that it might modify the
regulations. This generated excitement among ethicists and cautious
hope among scientists that the system's dysfunctions might be

identified and addressed. Scientists and scholars held meetings, wrote papers, and submitted thousands of comments on the proposed changes. OHRP announced the final changes in 2018.

Two of the changes are clear improvements. One exempts the type of study that Coe conducts that uses discarded biological materials. Every patient who comes to Coe's clinic leaves a urine sample. These samples are usually poured down the drain at the end of clinic, but sometimes Coe's team uses them, with the urine donor's identifying data removed, for protein analysis. The new regulations allow Coe, at least in theory, to conduct this analysis without specific consent. In another positive change, the regulations stipulate that approval from a single IRB will usually suffice for studies that are conducted at more than one institution. The other changes are mixed; when they all settle out, I suspect that most scientists, in most fields, will find their regulatory burden little changed. Coe tells me that his experience has not improved.

This process has been a triumph from OHRP's point of view, and not only because it has eliminated a few points of friction. More significantly, the protracted sequence of proposals and revisions kept the system's critics focused on issues identified by OHRP and sucked the oxygen out of any discussion of more fundamental reform. OHRP's ground rules assumed that the only goal of the IRB system is to protect prospective subjects from harm, and that progress, meaning public benefit from biomedical research, remains optional.

OHRP assumes as well that we should retain the changes that have happened over the decades—the stripping of the authority to reverse an IRB decision from university officials, the increasing obsession with risk, the commitment to increasingly elaborate consent forms. In short, it assumes that the current system, using a single set of rules administered by a single federal office, focusing on protection and ignoring the value of research, works well and could work even better. During the years of discussion, few attempted to compare the system's successes and failures. Few asked whether we would do better if we returned to something like the system's original methods and goals.

This alliance of ethicists and regulators has largely ignored the criticisms of individual scientists like Coe and of organizations like the American Association of University Professors. Even the National Academy of Medicine has fared no better, since in 2009 it documented the damage done to research by federal privacy regulations (which are often implemented by IRBs) and recommended they be substantially modified. This plea was met with silence.

THERE IS DEEP-SEATED RESISTANCE to fundamental change among the people who make the IRB system work, the committee members and staff, and among those academics who have made it their life's work. They concede the system's faults, but believe that it is not only the best way to protect subjects but the only way to prevent a repetition of the Tuskegee syphilis experiment. In their view, we must preserve the system so that racial and ethnic minorities may continue to continue to enjoy its protection. This belief is encouraged by the government, and not only OHRP. As I write this, the CDC website says, "After the Tuskegee Study, the government changed its research practices to prevent a repeat of the mistakes made in Tuskegee." This conviction that the IRB system protects racial and ethnic minorities is both durable and unsubstantiated. I know of no published or unpublished evidence that IRBs are actually identifying and blocking research proposals that would injure minorities.

Ironically, the commitment to preventing "another Tuskegee" may now be contributing to a different error that also victimizes members of racial minorities. Scientists once thought there was no problem when white men were the only subjects in the large trials that develop new treatments. We now recognize that excluding women and minorities from these studies damages the generalizability of their results. When Vivian Pinn was president of the National Medical Association, she stressed the importance of including minority subjects in new drug testing as one way to reduce health disparities.

It is harmful when large studies, like the SUPPORT trial, are so tightly controlled that minorities are disproportionately excluded. As professor of social medicine Giselle Corbie-Smith writes, protection with benevolent intentions may still lead to "the problem of 'too much protection.'" Journalist and historian Harriet Washington, after describing abusive experiments in detail, pivots to ask us to "hold two seemingly contradictory but actually complementary facts in mind": African Americans must remain wary of research abuses, but they must also "welcome and embark upon medical research as a bridge to fording the gulf that yawns" between Black and white.

That gulf includes lung cancer, which kills African Americans disproportionately. This is particularly true in inner cities where single cigarette sales (illegal under federal law) are a cheap way for a teenager to become addicted to nicotine. Ruth Malone, of the School of Nursing at the University of California, San Francisco, planned a study of the local sale of these cigarettes. The IRB, for reasons that were never fully clarified, appears to have been afraid that the study would be controversial. Whatever the cause of its concern, its response to Malone was a firm "No." She had engaged local residents as co-investigators, and they were as baffled as she was. As one of them said, "It's like Tuskegee in reverse."

We should all remember the Tuskegee syphilis study and heed its lessons. But we should not allow fear of the past to keep us from planning for a better future.

PART OF THE PROBLEM WITH THE IRB SYSTEM is the attempt to use ethics to organize and run a government regulatory program. This is not to say that government action should be unethical or amoral.

Every government program should be conducted ethically. Every program should promote goals that can be seen as having a moral component, such as security, justice, and equality. The Department of Justice even has a moral concept in its name. But formal ethics, the kind taught in an academic department of philosophy, had not previously been used to help write regulations and guide their implementation. It was an experiment, and the results are now in.

Congress invited the philosophers in as part of the National Research Act of 1974, asking them to help draft the regulations and shape the way those regulations would be interpreted. From that time to this, philosophers, and those trained to follow philosophical methods, have argued that their disciplined and sophisticated thinking is superior to common-sense notions of right and wrong in research oversight. They have no choice, for if they are to make a contribution, their recommendations must sometimes contradict our everyday intuitions.

Seizing the high ground gives the philosophers many advantages. Scientists who are confronted by what seem like ridiculous IRB policies have no easy way to rebut the claim that "ethics" requires a particular change. Casting research oversight in terms of right and wrong allows the system's defenders to wonder out loud if its critics have lost their moral moorings. Speaking in the name of ethics gives the philosophers moral power.

But with that power, philosophy should accept a measure of responsibility when things do not go as planned. Sociologist Laura Stark writes of "a system that both investigators and federal regulators agree is broken." Greg Koski, who was the first head of OHRP, writes that "a complete redesign of the approach, a disruptive transformation, is necessary and long overdue."

Some ethicists, to their credit, have done their best to diagnose the IRB system's worsening dysfunction and propose solutions. In 2005, Rosamond Rhodes, of the Icahn School of Medicine at Mount Sinai in New York, wrote a powerful attack on the "conceptual missteps" and "accepted dogmas" of the field and called for a reconsideration from the ground up. A few years later, she proposed that research with only negligible hazard be subjected to lower standards of consent. In 2013, Ruth Faden, of the Johns Hopkins Berman Institute of Bioethics, proposed that research that is closely tied to clinical care should be evaluated using a completely new ethics framework. Ethicist Ezekiel Emanuel and coauthors reviewed the IRB system's difficulties and identified fifteen specific problems. They believed that a solution required "five fundamental reforms."

These are just a few of the many suggestions for change. Koski takes a more pessimistic view, writing that trying "to regulate the implementation of ethical principles of conduct, is basically a failure."

This failure is unsurprising. Plato dreamed of the wise rule of a philosopher king, but until the IRB experiment, philosophers had not claimed expertise in the nuts and bolts of government operations. Now that the question has become how to fix a regulatory system that is itself sick, the Department of Philosophy is unable to agree on a diagnosis or treatment.

WE CAN SEEK A SECOND OPINION. To obtain it, we need only leave the Department of Philosophy and find our way across our virtual campus to the Business School, a journey in both location and method. Business school scholars like to connect general principles with real-world problems, and they often collaborate with commercial consultants and leaders from business, education, and government in their search for effective interventions. I suspect that some of the philosophy faculty feel that their own work is more elevated than that of their business school colleagues—that the business school is a little tacky. They are entitled to their opinion, but if the philosophers had lived up to their promises, we wouldn't be looking for help from people who are interested in how the world actually works. Who are interested, for instance, in how to manage risk.

When ethicists refer to risk management at all, they generally see it as a shameful and inappropriate attempt at self-protection. Ethicist Michael McDonald and colleagues ask if "one might ... wonder whether [IRB systems are] used more for risk management by research institutions and sponsors rather than for genuine protection of research participants."

There is no need to wonder! IRBs today are in no doubt that their primary mission is to protect their institutions from federal punishment. But when an IRB, acting under misguided federal supervision, privileges institutional protection over the welfare of research subjects and society at large, it is not practicing risk management. It is just covering its ass.

This reflexive self-protection is a proof of the system's dysfunction, not an argument that IRBs should not be doing risk management at all. After all, the system's central purpose is to manage the risk that subjects encounter in research. Further, the interests of subject and institution have substantial overlap. In a properly-structured system, any committee decision that protects subjects also protects the institution.

Risk management is emphatically not the suppression of risk by any means available. It is also not trying to shift the blame after a disaster—that's public relations. As our colleagues in the business school will gladly explain, risk management is a term of art that refers to a specific set of methods developed, step by step, over the last fifty years, and whose goal is to prevent the disaster in the first place.

Insurance is the grandfather of risk management, and for centuries it was the only formal method available. Over the generations, insurance policies have protected countless individuals and organizations from devastating financial losses. Nevertheless, institutions continued to suffer losses from hazards that could not be insured. By the 1960s, forward-thinking institutions were delegating identified uninsurable risks to a specific person or committee. This apparently sensible approach regularly led to unintended consequences as institutions learned (surely not for the first time) that someone who is ordered to reduce a specific peril may act in ways that thwart the organization's overall goals.

Too often, the result of suppressing a particularly worrisome risk was disaster that struck from an unexpected direction—a classic case of unintended consequences. It is perilous to establish a system that manages risk in a silo, without an awareness of the institution's overall goals. Corporations were not the only victims. Churches, foreign governments, and NASA all failed to manage risk well, leading to sexual abuse scandals, the explosion of a Cypriot naval base, and the fiery death of astronauts.

As business school scholars examined these disparate disasters, they realized that all risks are not the same. Some promise no benefit whatsoever. These pure risks can only result in loss, and they should be eliminated insofar as possible. Stanford Medical School, for instance, rests on the San Andreas fault. The librarian can bolt bookshelves to the walls without impairing use of the books. This is one way to reduce seismic hazard, which is a pure risk.

Other risks, however, are an inescapable part of the institution's operations. Stanford cannot provide medical care without some risk to patients; it cannot conduct research without some risk to subjects. Business school theorists label these opportunity risks, and opportunity risks are part of life for every organization. A church, for example, takes opportunity risks when it moves to a larger building. Will its congregation grow to fill (and pay for) the new space? Will there be unanticipated problems with financing or insurance? Similarly, a nonprofit takes opportunity risks when it hopes that a 5K walk/run for breast cancer research will earn more than it costs. Both organizations must decide, after balancing the risks and benefits, if it makes sense to proceed. In these go/no go cases, a decision obviously needs to be made; but even in routine operations, new hazards may expose hidden flaws in decisions made years ago.

Risk management theory provides a set of tested methods, embraced by institutions of every kind, for handling opportunity risk. Yale University and the US Navy both provide risk management services in-house, as does the Church of Jesus Christ of Latter Day Saints, otherwise known as the Mormons. The Quakers obtain risk management services from a variety of outside companies.

Some organizations prefer to keep these activities confidential, but NASA shares its comprehensive risk management methods online. It faces incredibly diverse risks: of injury to astronauts, equipment failures, budget and timetable overruns, and much more. As part of the agency's process—and this approach is appropriate for every institution—it focuses its effort on the risks that are both serious and likely. After all, there is never enough time and money to block every conceivable risk, nor is it wise to do so. As Justice Stephen Breyer observes, writing about attempts to ensure the public health, "the regulation of small risks can produce inconsistent results, for it can cause more harm to health than it prevents." Major risks must be addressed specifically; minor risks are often better handled by the system's informal operations than by dedicating resources to them.

Biomedical experimentation poses risks of all magnitudes, including serious risks that require thoughtful attention. In standard risk management theory, risks of this kind should usually be managed by a committee. This committee reports to, and may be reversed by, senior officials who are responsible for the overall success of the organization. This general framework is implemented in different ways that reflect the particular circumstances of each organization. Finally, because all things change with time, risk management theory dictates that this structure and process be flexible.

Risk management theory had not yet been developed in the 1950s, but James Shannon designed a system very much along these lines at the NIH's own Clinical Research Center. The CRC reviewed only studies that posed significant risks. Scientists could appeal to Shannon, who could, but apparently never did, reverse its decision. And the system was utterly flexible, for Shannon could modify it at will.

In 1966, it may have seemed easy to transplant the NIH's method to other institutions. The model was simplicity itself—a committee ensured that risks to subjects were carefully reviewed and that the consent process was appropriate. Shannon could not have anticipated that his policy requiring other institutions to follow the same model would become a point-by-point illustration of the painful lessons that inform risk managers today.

Review at the NIH had not been conducted in a silo, since every member of Shannon's committee was an institutional leader. While its sole charge was to protect research subjects' welfare and rights, committee members understood where subject risk fit into the institution's overall goals. Subject protection was the reason they met, but the committee was also well aware of NIH's mission of developing innovations that would save lives.

In 1966, when research institutions everywhere were required to adopt the model, it was often implemented in parallel with the original structure at the NIH. Case Western's first IRB, for instance, was led by the dean of the school of medicine, and the members were the departmental chairs. Like any IRB, this committee scrutinized the risks and consent processes of proposed research, but its members were also responsible for the institution's operations at the highest level. This structure prevented siloed thinking.

NIH's Clinical Research Committee had considered only protocols that seemed worth reviewing, but federal overseers soon insisted that the newly-formed university and medical school IRBs scrutinize with equal care every federally-funded research proposal, no matter how low its risk. The flood of routine protocols that followed led to predictable results. By 1974, the Dean at Case Western had yielded his place in the committee to the hospital chief of staff. The department chairs, too, discovered that IRB service demanded more time than they could spare—and probably less satisfaction than they expected—and they also stepped aside in favor of lower-ranking faculty. The initial tight connection between the IRB and the institution's leadership was broken as committee members were increasingly drawn from the ranks of junior faculty.

This change in membership created committees that were vulnerable to siloed thinking, but it did not condemn the system to its present dysfunction. Institutions that manage risks well routinely protect subordinate committees from being lost in their silos by appropriate oversight. Risk management theory is clear on this point: Final responsibility for all risks (and opportunities) resides at the highest level of the institution, usually the board of directors, which vests operational authority in the president and other senior officials, who in turn delegate specific responsibilities down the line. Staff (individuals or committees) at any level identify risks and take the action they feel is appropriate. If other people object, they may appeal to the senior leadership, which is responsible for adjudicating disagreements at lower levels and, more generally, integrating information about risks and opportunities.

Today, this sensible arrangement is virtually universal except in the idiosyncratic world of IRBs. When an IRB blocks or modifies a study, the regulations forbid institutional leadership to reverse that decision on appeal, no matter how ill-informed or poorly reasoned it may be.

Risk management theory identifies one more problem with the current system. The IRB system at its creation could be modified by a stroke of an official pen. It became more difficult to modify the system when it was embedded in the Code of Federal Regulations, and more difficult still with the passage of the National Research Act.

It doesn't have to be this way. Congress can recognize the system's many problems and create a better one, described more specifically in the next chapter, which is modeled on the NIH's Clinical Research Committee and modified to use contemporary risk management methods.

IN 2016, I SUBMITTED A PAPER to the journal *Pediatrics* about several proposed studies that were thwarted by inappropriate IRB restrictions. One of the peer reviewers who recommended that my paper not be published wrote that "in each of these situations, there are studies that have gone forward, with IRB approval. ... the anecdotal horror stories of inappropriate IRB restrictions do not tell the whole story—and since it seems unlikely that IRBs will be abolished anytime soon, a paper might better focus on ways to conduct such research under current regulations than on examples of situations in which such research could not be conducted."

I'm happy to stipulate that much research proceeds without problems, that many scientists support the IRB system (and serve on their IRBs), and that this book is not comprehensive. It reflects one person's experience and opinions. I leave it to my scientist readers to judge how well it matches their own reality. I leave it to my lay readers to judge the persuasiveness of its arguments. And I leave it to Congress to take action, perhaps as outlined in the next chapter.

# 11

# Reform

C ONGRESS PASSED THE LAW that governs the IRB system. Congress introduced ethics into the system's culture and regulations. Congress threatened to sack Gary Ellis because it felt his agency should be more aggressive. Now is the time for Congress to create an oversight system that will protect subjects without handcuffing scientists. In this final chapter, we will consider the specifics of that much-needed change.

The first step is to identify the factors that must be balanced in a reformed system, and this requires a subtle pivot. Much of this book has spotlighted how IRB oversight burdens scientists. Fredric Coe, for instance, complains that IRB review costs him time, wastes his money, and impedes the training of younger doctors in his department. This emphasis on investigators' burdens might suggest that research oversight should balance the welfare of subjects and the right of scientists to conduct their work.

I have watched Coe's patients as they seek his counsel, and while their faith in him, and their affection for him, are obvious, I believe they would frame the matter differently. No doubt they would be sympathetic about Coe's frustrations, but they would care more about how IRB oversight deprives them (and the fourteen million other Americans who have had kidney stones) of the benefits of his research. I do not mean to minimize the importance of the scientists' complaints—an oversight system that arouses the animosity of thousands of investigators is a serious problem. But when we consider the problems of IRB review, we should look beyond the cost paid by the scientists to the price paid by us all. The balance that any system of research oversight should strike is therefore between subject welfare and the societal benefit that good science brings.

That is why this book is not about making life easier for scientists. It is about freeing the scientists to make life better for us.

Regulation helps keep us safe when we drink a glass of water, build a home, climb into a car, or go to the emergency room. It is an essential part of modern life. In the IRB context, formal oversight of biomedical research will always be essential. The critical question is how that oversight should be structured.

Before Congress begins the process of reform, it needs to understand that it has been a mistake to encourage OHRP to act like the police. Other changes, as I outline in the rest of this chapter, will be needed as well. Risk needs to be seen not as something to be defeated but as something to be considered thoughtfully. Consent needs to be seen not as something intended solely to protect potential subjects—a function it does not do very well—but as a process in which potential subjects, scientists, and institutions all have a stake. Then there is the question of flexibility: Research is conducted in an incredible variety of settings, and each institution should have substantial independence in how it protects subjects without needlessly hampering progress. Finally, Congress will need to settle on the appropriate level of federal oversight for the entire process.

Reform of the system should begin with a return to its sensible beginnings. James Shannon relied on the Clinical Research Committee for oversight at the NIH's own hospital, and he believed it to be the best model when he rolled the IRB system out nationwide. What he did not know—could not know—was that the CRC had features that helped it work well, although they were invisible at the time. Those features were typical of what we today consider sound risk management. The expanded IRB system was stripped of these features and, without their protection, the system sank into dysfunction.

When Congress reforms the system, it should remember that the CRC was customized to fit the needs of the NIH's hospital; that universities, hospitals, state health departments, medical schools, military units, and biotechnology companies, all of which conduct research, differ one from another; and that each institution knows best how to manage the risks of research within its own walls. If the new system is not under the control of detailed regulations, each institution will be free to decide, based on its individual circumstances, what research should be reviewed, how that review should be conducted, how long permission to proceed should be valid, what records should be kept, and more.


THE REFORMED IRB SYSTEM will need to see risk in a new light. Until now, IRBs have seen experimental risk as the enemy, and research that posed a measurable risk has been considered presumptively unethical. This needs to change, since risk should be seen as something to be considered with care and managed wisely, not rejected out of hand. When IRBs permit informed subjects to accept a measure of risk, it is not only good for science, it respects potential subjects' autonomy.

Challenge testing for Covid vaccines is a good example of hazardous research that might be worth approving. Most vaccines are tested by taking a large number of subjects, giving the vaccine to half and a placebo to the other half, and waiting for people to get infected. Since the number of people infected is relatively small, vaccine trials are slow. Novavax, for instance, was proud to announce the efficacy of its Covid vaccine in July 2021, with 63 cases of infection in the placebo group and only 14 in those who received the vaccine (none severe). To obtain this result, Novavax enrolled 30,000 people.

The same proof of efficacy could have been achieved far faster by a challenge trial, in which subjects are given the vaccine or a placebo and then deliberately exposed to the virus. Such a study for Ebola could not be justified, since it kills about half of its victims. Covid is different, since its fatality rate is much lower, and it is lower yet among young adults. Early in the pandemic, before vaccines became available, the risk of death among those aged 75 to 84 was two hundred times the risk for those between 18 and 29. A challenge trial would exclude young people with diabetes, hypertension, or heart or kidney conditions, further reducing the risk of serious illness and death. The idea of the challenge trial is to test a vaccine in the healthy young to get a rough idea of its likely value in the old and infirm, and it has been forcefully promoted by Rutgers ethicist Nir Eyal.

Jeffrey Kahn is director of the Johns Hopkins Berman Institute of Bioethics. In November 2020, he and his coauthors argued that a Covid challenge trial should not be conducted, in part because such a trial would present serious ethical problems. They observed that, while young people are less likely to suffer serious problems from Covid infection, the risk is not zero, and the vaccine being tested might itself pose risks. There are, the group wrote, "many unknowns" that would make it impossible to accurately predict the risk of participation. This uncertainty "makes adequate disclosure next to impossible in the informed consent process."

The five authors of this report hold one MD, two JDs, and three PhDs, and they have extensive experience in ethics, so their opinion warrants respectful consideration. They are certainly correct that early in the pandemic, before any vaccine had been approved, there was no way to know the vaccine's side effects or the long-term risks of infection, and there were other uncertainties as well.

Yet much *was* known. Most importantly, it was known that the death toll in the young was in the thousands, while in the old it was in the hundreds of thousands. I think a reasonable young person might see this as a chance to take a real risk for the benefit of their grandparents and of the public at large. In my opinion, there are times when it makes sense to offer enrollment in research with real risks. The Covid vaccine challenge trial is a case in point.

Kahn and his coauthors disagreed, justifying their position by the uncertainty of the risks involved. Yet major life decisions always involve an element of uncertainty. No high school student makes a fully-informed decision about which college to attend, and even couples who have happily cohabited for years sometimes find that, when they marry, things are no longer the same. This might be for better or for worse, but the point is that major decisions are never made with the luxury of complete knowledge.

Kahn's group has an average age of sixty, and I am uncomfortable with them telling people in their twenties what risks they may take. If a young potential subject decides that the pandemic crisis justifies taking a risk, that decision is not only reasonable, but laudable. When we allow them to accept a measure of risk, we respect their right to choose. Even philosopher Hans Jonas might approve, for despite his intense concern with protection of the individual, he recognized that "otherwise inviolable prohibitions and taboos" may need to be lifted in a state of emergency.

Although potential subjects have a right to accept some risk, I think the level should be capped. Some people are very risk tolerant, as the popularity of rock climbing and white water rafting attest, and some would no doubt accept substantial experimental risks. But in ordinary circumstances the IRBs of the future should not permit research in which serious injury or death is likely. This will to some extent restrict the autonomous choice of potential subjects, but science can make adequate progress against cancer and heart disease without offending the memory of Henry Beecher by allowing people to volunteer as martyrs.

This leads to another issue that American research institutions and government agencies pretend doesn't exist—the need for care of subjects who become sick or are injured in an experiment. The regulations require that subjects be informed as to whether they will receive compensation or medical care if they are injured as the result of an experiment. It does *not* require that they be given any compensation or care. Scientists have in the past purchased insurance for harm suffered during research. Experts and commissions have periodically recommended that this practice be resumed, and workable models in other areas, including no-fault workers' compensation systems, could be adapted to the research context. Almost every other country that is a major research sponsor already has such a system in place.

A REFORMED SYSTEM should also rethink consent—when it should be required, and what information potential subjects should receive. In terms of the latter, I hope that consent forms will contain accurate information, and no longer exaggerate risks and discount benefits.

The present system's requirement of a single template that specifies the elements of consent, sometimes modified if special circumstances require, is ill-suited to the limitless variety of research. With many different kinds of studies, the consent process should differ as well, so generalities will only get us so far. Still we can consider some of the opportunities presented by a new system of oversight.

Most medical care is provided on the basis of informal or implied consent. If a nurse approaches with a blood pressure cuff, I provide consent by rolling up my sleeve. Some research should proceed on the same basis, or, if written consent is required, it should be minimal. Fredric Coe's patients should be able to donate a cup of urine without signing a multipage form.

For research that is essentially zero risk, as when Coe analyzes urine samples or reviews medical records, consent forms offer little that most subjects will value. Under today's system, these forms explain what Coe will do, why he will do it, the risks that are anticipated (chiefly if confidentiality is breached), the medical benefits that the subject may expect from participation (always zero), and considerable boilerplate. His patients assume that any value in the form is not worth the time it takes to read it. When his IRB insists that he prepare such a form, and that his subjects read it, the trend toward meaningless consent is only heightened. It's the same as when we want to download a program online and, ignoring whatever information is presented, scan the screen looking only for the box labeled "I agree to the terms and conditions" in order to continue.

In the experimental context, formal consent is appropriate when research involves significant burdens or hazards. That is, after all, the point; it makes sense to seek consent only when there is a burden to accept or a risk to agree to. A scientist who plans to study the impact of sleeplessness on thyroid function, for instance, should alert prospective subjects that they will need to stay up all night. When there are no burdens apart from the obvious, such as feeling tired the next day, and when hazards are trivial or entirely nonexistent, formal consent is absurd and should be omitted.

Consent should also satisfy subjects' desire to know how the study may help others. IRBs today sometimes prohibit statements about the potential benefits of research on the theory that they are coercive. This misunderstands the meaning of the word "coercion," since by definition it requires force or the threat of force. IRBs' fear of describing the potential benefits of a study leads to a cautious description, carefully stripped of optimism, that can leave potential subjects baffled as to why the research is being done. Psychologist and ethicist Jonathan Baron speculates—and I think he must be right—that subjects who are given no explanation for why research is being done are likely to "insert their own hunches" and thus "make decisions on the basis of random beliefs that have little relation to the best information they could get."

Cumbersome consent procedures have had the unintended effect of reducing the enrollment of vulnerable groups, as for instance in the SUPPORT study of infant oxygen. This is problematic, since research in a homogeneous study sample cannot be relied upon in treating diverse patients. We should avoid complex consent forms or processes that reduce subject diversity and bias study results.

Commentators have long disagreed on the primary goal of consent: Is it to protect the institution, to protect the subject, or to enable the subject to make an autonomous choice? Actually, it is a mistake to view these goals as exclusive, since all three are important. I hope that a reformed oversight system will see the consent process as a negotiation in which both subject and institution have something to gain. We should admit what is obvious—that institutions use the consent form to protect themselves and promote their interests, as they have been doing for more than a century. We can also see that subjects, with IRB assistance, use consent to protect themselves, promote their own interests, and make autonomous choices.

The consent form is not a legal contract, but there are instructive parallels. In a contract, both parties agree to the terms, and both parties get something of value. If one party gets nothing, there is, by definition, no contract. That is why a parent will sometimes sell a home to a child for one dollar—the nominal payment makes the contract valid.

In the research context, with the signed consent form on record, the institution and its scientists can proceed without worrying that the subject will later deny agreeing to participate or being informed of the study's risks. The subject, too, gets something, whether cash, the possibility of medical benefit, or social reward (either as praise from others or the inherent satisfaction of altruism).

Consent to participate in research is not an agreement between equals, since negotiations between scientist and subject are necessarily one-sided. The scientist will always know more about the study than the subject, who may also be sick or stressed, and the scientist may be tempted to soft-pedal the risks in order to expedite the process. Today's system errs by permitting indigestible forms and the exaggeration of risks and minimization of benefits, but it is right in attempting to use consent to protect the rights and welfare of potential subjects. Tomorrow's system should pursue the same goal through better means.

In the new system, IRBs would be free to adapt sensible consent forms. Just what such a form would include would differ between studies, but a general template might include a series of sections, beginning with information that is both relevant and easy to understand:

1. A summary of the central research question, written in the simplest possible language. This might be just a paragraph or two, and it would be understandable even by prospective subjects who are not skilled readers. The beginning of the form for the SUPPORT study, for instance, might include a statement that it aims to find out how much oxygen is best for very early babies like yours.

2. A fuller but still limited description, also in elementary language. For the SUPPORT study, this section might state that doctors agree that premature babies need extra oxygen but do not know just how much is best. Premature babies are at risk of complications. Too little oxygen increases their risk of brain damage or death. Too much oxygen makes blindness and lung problems more likely. These problems can occur in any premature baby, but the goal is to avoid them as much as possible. This study will give some babies a little more oxygen and some a little less in order to find out what level of oxygen is the best overall. Every baby will receive a reasonable amount of oxygen, and every baby will receive the best care the doctors and nurses can provide.

3. Everything relevant. This section is for potential subjects (or parents) who want to know all about the study, including a complete list of possible risks.

4. Even a reformed IRB system will still need to satisfy state laws, each institution's specific concerns, and other considerations that are of no interest to subjects. This information could be presented in the final section. Because it is irrelevant to subjects, it might be titled "Additional Information." This will tip off potential

subjects that they can skim or skip it without missing anything of value.

THE NEW SYSTEM should not be governed by a detailed set of regulations. Instead, IRBs should be free to use their experience, common sense, and guidelines developed by other institutions or specialty societies to decide which risks to permit and which to prohibit.

Even in the current dysfunctional system, it is striking how hard IRBs continue to work. We owe IRB members everywhere our thanks, and should rely on them as the heart of a reformed system, with the expectation that they will make reasonable choices when there are no right or wrong answers. In fact, we call on committees to make decisions of this kind precisely because there is no easy choice. Some committees will make these decisions wisely, some will not. But the reformed system does not have to be perfect, it only has to be better than what we have now. That won't be hard.

To make these ideas more concrete, we will consider how research oversight might be conducted under the reformed system by three Houston-area institutions.

NASA is one of the many institutions that have embraced modern risk management principles. Before a space mission ever nears the launch pad, the agency's engineers compile a list of everything that could go wrong (it's a long list). Trying to block all of these hazards would keep the agency from getting out of bed in the morning. Instead, its engineers go down the list and estimate how likely each risk is to materialize and how severe the consequences would be. This allows the agency to ignore risks that are unlikely or unimportant, and focus on those that pose genuine threats.

As part of its mission, NASA conducts medical research intended to protect astronauts from the immediate and long-term hazards of space travel. The most obvious is death by explosion. Those who serve on the International Space Station also face hazards of prolonged life in space, including blood clots, muscle atrophy, balance disorders, and bone loss. ISS astronauts also show urinary changes that predispose them to kidney stones. NASA is unlikely to send Fredric Coe into space to minister to ailing astronauts, so it does what it can to study the physiological changes that might lead to problems in orbit.

In one study, NASA scientist Scott Smith and his coworkers evaluated 42 astronauts before, during, and after long-duration missions to the ISS. They were particularly interested in bone loss, which had been a major problem in earlier studies, and urinary changes. They found that both sexes benefit from the modern bone protection program, which includes exercise and medication. They also found that men seemed to be more likely than women to develop urinary changes that can precede the development of kidney stones.

NASA has its own IRB, which follows the usual rules, and which had to approve Smith's request to obtain blood and urine samples and take bone density measurements. I do not know just how the IRB wanted Smith to explain the risks of these procedures to the astronauts, nor how their consent was memorialized, only that this IRB operates under the same rules as all the others.

Under a reformed system, NASA's IRB could create its own process for research oversight, based on its unique mission and resources, and it would be free to decide, for instance, that studies that involve only urine and blood collection and densitometer readings could be conducted by the doctors involved without meticulous IRB oversight.

Under a reformed system, every other institution would be equally free to manage the risks of research in its own fashion. The University of Texas MD Anderson Cancer Center in Houston treats thousands of patients each year and conducts research into every aspect of cancer causation, prevention, and treatment. It knows its own resources, its own challenges, the population that it serves, and the scientific questions it addresses, and under a reformed IRB system it would be responsible for establishing a system to manage the risks of its research.

As one example, MD Anderson's reformed IRB will need to address the special circumstance of patients with advanced cancer who are considering enrolling in a Phase 1 trial of a new chemotherapeutic drug. These patients hope that the drug will work for them, and although the chance of cure is vanishingly small, the chance of some benefit—some improvement in the quality of life, or extension of its duration—is significant. On the other hand, there is always the chance that the new drug will do nothing or even make things worse. As the new IRB evaluates risks and approves consent forms, it will need to balance the facts of the trial and the expectations of potential subjects in a way that will seem reasonable both at the time and in hindsight. With 21,000 employees, it can easily set up a research oversight process suited to its needs.

Houston Hospice, with just 185 employees, is tiny in comparison. Although many of its patients also suffer from cancer, the role of research in their care is quite different. You might think that it would be inappropriate to experiment with people so close to the end of life. If so, you would be in the good company of George Annas, ethicist and law professor. Annas writes that the terminally ill are "desperate and, therefore, too vulnerable" and would be "unable to distinguish research from treatment." He recommends that they should be "disqualified" from research.

Annas believes that the terminally ill would be desperate for a cure, but he provides no evidence in support of his opinion. Wendy Terry and her colleagues took the time to talk with hospice patients to get their view. Their interviewees were very ill indeed—almost one-quarter of the people she interviewed died within the next 24 hours. This close to death, they felt they no longer had to please anybody and were candid about their convictions. They saw participating in research as a way to help others, and they felt that they themselves would benefit from making a contribution. They were insulted by the idea that they might be unable to make an informed choice about participation. They had come to terms with the end of their life, but were angry at the thought of being treated as if they were already dead. Many did have one major reservation about becoming research subjects: If participating in research might extend their lives, they would see that as an adverse effect and would be less likely to participate.

Houston Hospice knows its resources and limitations, and it knows the ways in which its patients struggle to meet life's final challenge. I would trust its judgment as to what risks its patients might want to accept and what consent they will find useful. I would be surprised if its decisions were identical to those made by its larger cousin, MD Anderson, or by NASA. Each will manage the risks of research according to its best judgment.

This was true for the original CRC, whose membership and operations were designed for its specific environment. It did not review routine, low-risk research, only experiments that carried significant risk or otherwise merited careful consideration. Rather than a complex set of rules, it relied on the judgment and experience of its members. Its decisions were subject to reversal by the senior leadership. And, most importantly, it was flexible and could be changed in response to changing conditions.

OHRP does not recognize the value of institutional flexibility. The precedent was set in the 1970s, when Ohio State University attempted to establish IRB subcommittees for each department. OPRR refused to permit this arrangement. The IRB system of tomorrow should emulate the flexibility of the original CRC.

FLEXIBILITY SHOULD BEGIN with the method of research oversight. Institutions like Houston Hospice are too small, and conduct too little research, to support an IRB along the current lines. Under the reformed system, they could adopt a simplified approach to oversee their occasional forays into research. They might, for instance, assign oversight to a single person.

Larger institutions, like MD Anderson and NASA, are likely to continue to vest primary responsibility in a committee. Shannon's 1966 policy specified almost nothing about the specifics of research oversight, but it did stipulate that it must use a committee. Between that day and this, IRBs have confirmed the wisdom of that requirement. A group that draws on the knowledge and experience of a group of people has a good chance of making smart decisions. I have seen this confirmed in my own experience serving on the Stanford IRB.

There are problems with how IRBs function today, to be sure, but the fact that they function at all is a tribute to the dedication of their (usually unpaid) members, who are willing to spend endless hours in labor whose rewards are slight. The IRB system needs fundamental reform, but its committees do not. In any case, IRBs will need to maintain their current form so long as they review some research, such as investigations conducted under FDA supervision, that requires it.

I also like the current requirement that IRBs include members who are not scientists and members with no tie to the parent institution. The regulations do not explain why these people are important, but it seems to me that their chief purpose is to remind the scientists on the committee that their decisions need to make sense to the outside world. If the committee is working well, they will never need to point out that a protocol under consideration would shock the public if approved. Their mere presence will keep such protocols from ever being seriously considered.

Yet committees can work no better than their circumstances permit, and IRBs, working in today's dysfunctional environment, have come up with some truly amazing nonsense. The IRB at the University of Colorado is composed mostly of faculty members. As we saw in chapter four, when Rob Knight wanted to study the skin microbiome, this distinguished group managed to identify both AIDS and smallpox as experimental risks. I am sure that, acting alone, no single member could reach this level of crazy. It took a village. The committee, however, was not the problem. The problem was in the rules it followed and the supervision it received from OHRP.

The first step in reforming the rules is to make it plain that IRBs should consider both subjects and society. Subjects' concerns extend beyond safety to include their autonomous right to choose a measure of risk. The relationship among the three elements involved—subject safety, subject autonomy, and society's need for better medical interventions—is complicated. It is not a zero-sum game, for science can advance without subjects being subjected to unacceptable risks. That win-win outcome is the one that IRBs and scientists should strive for.

WHILE THE CURRENT SYSTEM, in which officials in Washington micromanage every aspect of institutional risk management, is unworkable, I do not think federal oversight should be abandoned entirely. When Congress creates a new system, it should therefore establish some level of overall control of the system. Congress might choose to retain something like the current approach, with a dedicated agency endowed with the ability to impose substantial penalties. Alternatively, federal oversight could return to the approach in place in 1966, with a small office and limited enforcement power.

The argument for retaining strict federal control is simple: Scientists can be arrogant, institutions can be lax, and research can be abusive. For these reasons, it would be reasonable to retain tight federal control.

On the other hand, the failures of tight control have become obvious during the present system's tenure, for federal oversight has been strict but not wise. Over the last half century, OPRR and OHRP have deprived institutions of millions of dollars in federal funding for no good reason, hobbled the work of scholars and scientists whose work is incapable of causing the slightest harm, damaged and delayed major studies, and deceptively persuaded parents that their children's participation in research was a tragic mistake. After this humbling succession of failures, Congress could decide to design a new system that guides with a lighter touch. This would also be reasonable.

In either case, Congress should discard today's culture of punishment and recognize the need for balance among subject risk, subject autonomy, and public benefit. The scandals of the past should not be forgotten, but they need not hold the future hostage. And whatever Congress decides, I hope it will build a system designed to be revised as experience suggests.

The current system prohibits most appeals within the institution. This doctrine of IRB infallibility is one of the most pernicious aspects of the present approach. IRBs, like other committees, are vulnerable to fads, their members may have idiosyncratic views about research, and they may simply err. In a typical medical school, I expect that the new IRB will be subject to review by the associate dean for research, who in turn reports to the president and the board of trustees. Under this new system, Fredric Coe could appeal his IRBs' rulings to a senior institutional official.

JAMES SHANNON'S IRB SYSTEM, as established in 1966 and solidified by law in 1974, was an experiment, as are all attempts to manage our complex and changing society. Congress should try again, but it need not do so blindly. The present system's vicissitudes make apparent some traps to avoid, while advances in public policy and risk management suggest a better approach. No system will be perfect, but we can do better, and doing so will protect subjects' rights and welfare while improving the life of the nation, and the world.

# Acknowledgments

THIS BOOK HAS BEEN FIFTEEN YEARS in the making, and I am grateful to the many people who made it possible. I'm going to start with three of the four top debts.

To my benefactor, Regina O'Donnell. She established the William W. O'Donnell, MD, and Regina O'Donnell Chair in Family Medicine at Baylor College of Medicine and asked that I be named the first holder of that chair. Her generosity gave me the time which made it possible to write this book and her ongoing emotional support has helped in my darker hours.

To Carl E. Schneider, my invaluable friend, mentor, and collaborator. Carl and I spent years studying the IRB problem intensively. I found our conversations, and our debate, almost as physically pleasurable as helping ourselves to a sheet of fresh fudge brownies. When we began our work together, I thought I had a pretty good idea of the IRB system and the source of its problems. I soon learned how wrong I was. Carl was operating at a far superior level of understanding, one cut free of the invisible assumptions and blind spots that hobbled me. Thank you, Carl, for showing me so much. And thanks to Joan Schneider for welcoming me to your home through six years of summers and winters.

To Jon Tyson, neonatologist and medical scientist par excellence at McGovern Medical School in Houston. Jon has been a long-time critic of the IRB problem, and he sponsored a years-long discussion group dedicated to possible solutions, including paying part of my salary. The group was instrumental in developing many of the themes in this book. Paula Knudson, with a lifetime of experience in IRB circles, was always eager to teach and learn in these discussions, as were pediatricians Patricia Evans, Susan Wootton, and Kathleen Kennedy. We had significant guests as well, including Larry McCullough, Emil Freireich, Carl Elliott, and Ivor Pritchard.

My friends and family have put up with my obsession with IRBs since the mid-2000s. My first wife, Judy Levison, and I have shared more than twenty wonderful years together, two children, and our joint progression from community doctors to academics. Our children, Jordan and Diana, our son-in-law, Max Kaye, and my siblings and brother-in-law, Eunice Thomas, Roger Whitney, and John Thomas, did their best to keep me near sanity with mixed success. Eunice also read multiple drafts of this book and provided invaluable suggestions and comments. Shawna Peterson stood by me through good times and bad, and from Anchorage to Munich. Justice Norris has been a delightful traveling companion.

My connections to the publishing world go back to 1974, when I worked for Jane Isay at Yale University Press, and she was helpful as I worked on this manuscript almost half a century later. Freelance editors Susan Hatch Morgan, Kathleen Kearns, and Pamela Haag helped make the manuscript sharper and more fluid. Thomas McNamee provided helpful comments at every stage.

Getting this book into print has been a challenge. An early version of this book was under consideration at Harvard University Press. I considered Elizabeth Knoll my editor even though she eventually told me (in 2014) that the manuscript wasn't ready for publication. She was right, and she helped me understand why and how to do better the next time. A thoughtful rejection by an experienced editor is a true gift.

Once I had a new and improved manuscript, there followed a long search for a publisher. My first thought was of another university press. However, university press editors rely on peer reviewers to gauge the quality of submissions. This review system can have trouble distinguishing between a manuscript that correctly points out that an entire academic discipline is wrong and a manuscript (and author) that have gone off the deep end. Is this interesting manuscript, in other words, the work of a prophet or a crackpot? This manuscript, now the book you are holding, fell stubbornly on the wrong side of that divide. As an alternative to a university press, I considered trade publication, but the people I contacted felt that too few readers would be interested in the details of medical research and its oversight.

Just as I thought I would be stuck forever, I had the good luck to contact the wonderful Barbara Monteiro. In addition to serving as publicist for the book, Barbara connected me to Karl Weber of Rivertowns Books. Step by step, Karl has brought the manuscript through the stages of publication. He is the editor and publisher I needed, bringing his extensive experience and thoughtfulness to the table. His collaboration has made it possible to foreground the experiences of the scientists who are the heroes of this book and whose voices are too rarely heard. Our team was rounded out by proofreader Michelle Asakawa and by Barbara's husband, Len Maniace, an experienced science writer and editor.

I first attempted serious academic work at Stanford Law School and the Stanford Center for Biomedical Ethics. Henry Greely was my mentor and was the first to point out, not unkindly, that writing involves more than putting a series of sentences on paper, and to help as I encountered my own first conflict with the IRB. Ernlé Young and Tom Raffin, of the Center for Ethics, were very helpful. Shelley Correll, who was then a grad student at Stanford, helped me analyze the raw data for one of my early publications and kept me out of sex-and-gender errors. Service on the Stanford IRB, which was ably administered by Donna Adelman, and attendance with the Stanford group at meetings of PRIM&R, the national IRB organization, helped me understand IRB functioning both locally and as a national phenomenon.

Stanford Law School allowed me to take courses at the business school and in the department of epidemiology & population health, particularly biostatistics from Philip Lavori and the late and much-missed Byron "Bill" Brown. Bill was the only statistics teacher that I ever saw flip a coin in a classroom and have it come up heads four times in a row (he was making several points at once).

I have been very fortunate at Baylor College of Medicine in Houston. Vice Chair for Research Bob Volk helped ease me into a mixed research, teaching, and patient care position. Department chairs Steven Spann and Roger Zoorob were supportive even though they knew I was fighting a thoroughly established system. Department administrators Atma Ram and Sara Rahman provided help I much appreciate. Val Pavlik, Sue Nash, Jason Salemi, and Hamisu Salihu were always ready with sound insights. My colleague David J. Hyman never met a conventional wisdom he couldn't puncture ("So, you think you understand hypertension?"). He left us too soon but we remember his passion for the truth. David Buck was a lifesaver at multiple moments of crisis. Other friends, colleagues, and patients from my Baylor years who provided support and read various drafts include Joe Barnes, Andrew Wilking, Sarah Glass, Jeffrey Steinbauer, Jennifer Pratt Mead, and Bill Emmons.

Amy McGuire, who leads Baylor's Center for Medical Ethics and Health Policy at Baylor, has been a pleasure to work for and with. Baruch Brody, Ernest Frugé, and Larry McCullough were among the ethics center faculty who were helpful as I worked to keep the theory of ethics connected to the practice of medicine. Stacey Berg, another of my coauthors, was not only on the Ethics faculty—she ran our IRB. I hasten to add that this book is not written with the Baylor IRB in mind.

Any book benefits from the critical insights of other scholars. This book was enriched by the work I did with Carl Schneider on his own critique of the IRB system, *The Censor's Hand*. Carl recognizes the value of collaboration, and in 2011 he organized a research roundtable at the Searle Center at Northwestern Law School to review the *Censor's Hand* manuscript as it was then. Participants included Sydney Halpern, Carol Heimer, David Hyman, James Lindgren, and Elizabeth Knoll. Later that same year we held a second roundtable at the home of my sister and brother-in-law, Eunice and John Thomas. The lively discussion featured Zachary Schrag, Philip Hamburger, Jon Tyson, Susan Wootton, and Mark Hall, in addition to Carl and myself.

Speaking at conferences provides scholars with an opportunity to share their vision with a larger audience and discover new avenues of thought. Joan Rachlin invited me to participate in a plenary panel in 2007 at PRIM&R. Jon Hafner suggested that I have something resembling a debate with Ivor Pritchard, then acting head of OHRP, at an IRB conference at the University of Illinois College of Medicine at Peoria in 2008. It was nice to get to know Pritchard a little—he is a genuinely nice man—and I loved the slide deck for his talk, which was built around Judith Viorst's *Alexander and the Terrible, Horrible, No Good, Very Bad Day*. Mats Hansson brought me to Sweden, in 2010 and 2011, for major conferences at Uppsala University that featured international perspectives on the dilemmas of research ethics. In 2014 I gave a keynote address at the "Closing the Gap on the Most Deadly Diseases" conference at the University of Minnesota. In that same year, I enjoyed debating David Hunter over the cost in lives of ethics review at the World Conference of Bioethics in Mexico City.

Fighting the IRB system is made less lonely by a small group of scholars who are unafraid of the challenge. Carl Schneider was, for years, central to this group. Mats Hansson of Uppsala University is an inspiration. Philip Hamburger, of Columbia Law School, operates without regard for what others consider conventional limits on thinking. Will van den Hoonaard, a wise sociologist, has a humane vision of the field and is the author of *The Seduction of Ethics*, one of the foundational texts. Historian Zachary Schrag wrote *Ethical Imperialism*, the definitive account of the IRB system's overreach into the social sciences, and hosted an essential blog for years.

As a family doctor, I am ever anxious to be accurate when venturing into medical or scholarly fields that are not quite in my wheelhouse. Laura Stark, Rob Knight, Sean Berenholtz, Mark Hochhauser, and Richard Kallet each reviewed and, when needed, corrected sections within their areas of expertise.

I have many debts to other scientists. First and foremost to Fredric Coe, who with his colleagues Joan Parks and Arlene Chapman welcomed me to their kidney stone clinic and discussed the close relationship between research and practice. Rob Knight shared his story of AIDS and smallpox. Christina Hultman and Martin Ritzén helped me understand ethics review in the Nordic setting. Razelle Kurzrock, David Stewart, and Heidi Russell explored the challenges of IRB review in oncology. And it was at the end of a talk I gave on the IRB problem that Emil Freireich, one of the giants of American oncology, stood up to ask a question: "How can I help?" Thank you, Emil. We have lost you now, but every year your discoveries continue to save the lives of numberless children with leukemia. That's pretty good for a lifetime's work.

I have debts to other academics as well. Jerald Bachman of the Institute for Social Research at the University of Michigan gave me as much help as if I were working by his side in his lab, when in fact I was a law student in California who could give him nothing but gratitude and a bunch more questions. Kirsten Alcser of the Institute was equally helpful. Howard Brody welcomed me to the ethics fold when I was still a law student even though I had never met him, either. Barbara Evans of the University of Florida law school has a sharp eye, a wonderful way with words, and is perhaps the best storyteller in this list. Jonathan Baron sees ethics with an appropriate degree of skepticism. Pauline Rosenau and I discussed the mysteries of academics on more than one occasion. Stephen Ceci has been exposing the self-deceptions of the IRB system since 1985. Steven Pinker knows how important it is to speak out about what really matters—and in the case of the IRB system, that is letting the scientists get on with the work of saving lives.

These people, approximately seventy of my friends and colleagues over the years, have given me many insights and saved me from innumerable blunders, and I owe them my profound thanks. Needless to say, any errors that remain are entirely their fault.

I have saved my top debt for last. It is to my wife, Patricia Whitney. It was she who pointed out that what IRBs are trying to do is understood in any other setting as ordinary risk management, and thus provided the clue to the system's fundamental error and the path to its reform. Of equal importance was her unsupported conviction that I could write this book long before I believed it myself. My love and thanks are for you, Patty.

Boulder, Colorado
April 2023

# Source Notes

## 1. Unintended Consequences

7    **an injection of live cancer cells:** Jay Katz, Alexander Morgan Capron, and Eleanor Swift Glass, *Experimentation with Human Beings: the Authority of the Investigator, Subject, Professions, and State in the Human Experimentation Process* (New York: Russell Sage Foundation, 1972), pp. 9 ff.

7    **had a story of their own:** Rebecca Skloot, *The Immortal Life of Henrietta Lacks* (New York: Crown Publishers, 2010).

8    **that lost her to cancer:** Henry T. Greely, "Legal, Ethical, and Social Issues in Human Genome Research," *Annual Review of Anthropology* (1998): 473-502.

8    **had long abused minorities:** Harriet A. Washington, *Medical Apartheid: The Dark History of Medical Experimentation on Black Americans from Colonial Times to the Present* (New York: Doubleday, 2006).

8    **the Public Health Service:** James H. Jones, *Bad Blood: The Tuskegee Syphilis Experiment* (New York and London: Free Press, 1981).

9    **For more on these two systems:** At the federal level, two agencies, FDA and OHRP, use IRBs to protect research subjects. An

IRB can work under the supervision of one or of both (fortunately their rules are similar). The jurisdiction of FDA and OHRP sometimes overlaps. New drug, medical device, and biologic development takes place under FDA supervision. Research conducted or funded by the Department of Health and Human Services takes place under OHRP supervision. If there is NIH-funded research that is meant to develop a new drug (not a common event), then both FDA and OHRP have authority. Finally, many research institutions require all research, regardless of funding, to be reviewed by an IRB, which sweeps in a great range of unfunded and very limited projects.

9     **around the world:** Each country has addressed the moral and practical challenges of human experimentation in its own way. Denmark and the United Kingdom, for instance, do not conduct research oversight in just the same way as the United States. But the American influence is substantial. Much of the literature in research ethics comes from the United States, and, even more to the point, much of the funding for research comes from the US as well. This funding is sent abroad with a string: Local ethics review must be at least as stringent as that conducted in the US. As a result, many aspects of American oversight have been exported as well. While this book focuses on research in the US, the issues it raises apply internationally.

9     **delayed their development:** Robert Steinbrook, "How Best to Ventilate? Trial Design and Patient Safety in Studies of the Acute Respiratory Distress Syndrome," *New England Journal of Medicine* 348, no. 14 (2003): 1393-1401.

10    **The biggest ethical lapse:** Carol A. Heimer, "The Unstable Alliance of Law and Morality." In *Handbook of the Sociology of Morality*, 179-202: Springer, 2010.

10    **thousands of unnecessary deaths:** Simon N. Whitney and Carl E. Schneider, "Viewpoint: A Method to Estimate the Cost in Lives of Ethics Board Review of Biomedical Research," *Journal of Internal Medicine* 269, no. 4 (Apr, 2011): 396-402.

11    **Holocaust survivor Elie Wiesel:** George J. Annas and Michael A. Grodin, *The Nazi Doctors and the Nuremberg Code: Human Rights in Human Experimentation* (New York: Oxford University Press, 1992), p. ix.

11    **might refuse to participate:** Katz, *Experimentation with Human Beings*, pp. 9 ff.

12    **the modern era in research ethics:** Henry K. Beecher, "Ethics and Clinical Research," *The New England Journal of Medicine* 274, no. 24 (June 16, 1966): 1354-1360.

12    **he established a panel:** Laura Jeanine Morris Stark, *Behind Closed Doors: IRBs and the Making of Ethical Research* (Chicago; London: The University of Chicago Press, 2012).

12    **provided by their institutions:** Usually this was no oversight at all. See Irving Ladimer and Roger W. Newman, *Clinical Investigation in Medicine: Legal, Ethical, and Moral Aspects: an Anthology and Bibliography* (Boston: Law-Medicine Research Institute, Boston University, 1963).

12    **to break with this tradition:** Mark S. Frankel, "The Public Health Service Guidelines Governing Research Involving Human Subjects: An Analysis of the Policy-Making Process," PhD dissertation, George Washington University, 1972.

12    **in 1966, a federal directive:** Office of the Surgeon General, US Public Health Service. "Surgeon General's Directives on Human Experimentation." *American Psychologist* 22, (1967): 350-355.

13    **James Theodore was our IRB chair:** Details of IRB members, operations, and consent form requirements are in the Stanford University "Manual for Administrative Panel on Human Subjects in Medical Research," 1996-1997, in author's possession.

13    **The regulations encourage diversity:** The regulations governing IRB review of federally-funded medical research have been promulgated jointly by sixteen federal departments and agencies and can be found in the Code of Federal Regulations at 45 CFR 46, "Protection of Human Subjects." In its most recent revision, the Department of Health and Human Services took the lead role, and within HHS, the Office for Human Research Protections did most of the work.

15    **sometimes bought insurance:** Sydney A. Halpern, *Lesser Harms: The Morality of Risk in Medical Research* (Chicago: University of Chicago Press, 2004).

16    **as well as codeine:** Melinda Moir, E. Bair, P. Shinnick, and A. Messner, "Acetaminophen Versus Acetaminophen with Codeine After

Pediatric Tonsillectomy," *The Laryngoscope* 110, no. 11 (2000): 1824-7.

17    **Baylor College of Medicine:** We have no connection with Baylor University in Waco, Texas. The University and Medical School parted ways in 1965.

18    **a former federal official:** The official is Greg Koski, who was the first head of the Office for Human Research Protections. Greg Koski, "Research Ethics and Oversight: Revolution, Or just Going Around in Circles?" *The Monitor* 21, no. 55 (2007): 2007-2057.

18    **To the dismay of the research community:** Although IRBs are generally doing their best in a difficult situation, they could be doing more. I have written a manual providing the details; see Simon N. Whitney, *Balanced Ethics Review: A Guide for Institutional Review Board Members* (Cham, Switzerland: Springer, 2016).

18    **This book grew out of that frustration:** This book is one of several recent book-length explorations of the IRB world. Sociologist Laura Stark, sociologist Sarah Babb, and psychiatrist Robert Klitzman have each written books based on interviews with those on the front lines of IRB operations. Historian Zachary Schrag's book examines the fraught relationship between IRBs and social sciences. And legal scholar Carl Schneider has written a vigorous attack on the system and the horse it rode in on. See Stark, *Behind Closed Doors*; Sarah L. Babb, *Regulating Human Research: IRBs from Peer Review to Compliance Bureaucracy* (Stanford, California: Stanford University Press, 2020); Robert L. Klitzman, *The Ethics Police?: The Struggle to make Human Research Safe* (Oxford; New York: Oxford University Press, 2015); Zachary Schrag, *Ethical Imperialism: Institutional Review Boards and the Social Sciences, 1965-2009* (Baltimore: Johns Hopkins University Press, 2010); and Carl E. Schneider, *The Censor's Hand: The Misregulation of Human-Subject Research* (Cambridge, MA: MIT Press, 2015). I worked with Carl on early drafts of this book but he and I went our separate ways before it was completed. This list is incomplete; other excellent reviews and commentaries include Nancy M.P. King, Gail Henderson and Jane Stein, *Beyond Regulations: Ethics in Human Subjects Research* (Chapel Hill, NC: University of North Carolina Press, 1999) and Halpern, *Lesser Harms*.

18 **the University of Chicago kidney stone clinic:** Many thanks to Fredric Coe, Joan Parks, Arlene Chapman, and their colleagues for their hospitality when I visited in 2017.

19 **in the *Lancet* in 1973:** Fredric L. Coe and Lynn Raisen, "Allopurinol Treatment of Uric-Acid Disorders in Calcium-Stone Formers." *Lancet* 301, no. 7795 (1973), 129-131.

19 **They are still standard:** UpToDate, "Uric Acid Nephrolithiasis," https://www.uptodate.com/contents/uric-acid-nephrolithiasis?search=uric%20acid%20kidney%20stones&source=search_result&selectedTitle=1~150&usage_type=default&display_rank=1#H7, accessed August 18, 2020.

20 **are regularly piled on the old:** Some of Coe's research is now reviewed by a commercial (for-profit) IRB. For an interesting discussion of the growth of commercial IRBs, see Sarah Babb, "The Privatization of Human Research Ethics: An American Story," *European Journal for the History of Medicine and Health* 78, no. 2 (2021): 392-411.

20 **Coe would be less vexed:** Fredric L. Coe, "Costs and Benefits of a Well-Intended Parasite: A Witness and Reporter on the IRB Phenomenon," *Northwestern University Law Review* 101, (2007): 723-733.

20 **A federal investigation:** Robert Steinbrook, "Protecting Research Subjects—The Crisis at Johns Hopkins," *The New England Journal of Medicine* 346, no. 9 (Feb 28, 2002): 716-720.

20 **to stop the experiment:** That is the official view; others disagree. A scientist at Johns Hopkins told me that a literature review would not have raised alarms. Email from Edward Fuchs, June 7, 2016.

20 **the Chicago IRB requires:** The IRB's instructions are at University of Chicago, Biological Sciences Division, IRB Frequently Asked Questions (FAQs) and Submission Guidance, https://biologicalsciences.uchicago.edu/irb-faqs-and-guidance, copyright 2020, accessed 5/16/20.

21 **presumption of infallibility:** A higher institutional official can overturn an IRB decision to approve a proposal, but cannot change a decision to modify or block it. 46 CFR 112.

21 **become addicts:** Jonathan E. Shoag, Neal Patel, Lina Posada, Joshua A. Halpern, Talia Stark, Jim C. Hu, Brian H. Eisner, and

Jonathan E. Shoag, "Kidney Stones and Risk of Narcotic Use," *The Journal of Urology* 202, no. 1 (2019): 114-118.

22      **Flecainide, for instance:** We will review this in detail in chapter eight.

23      **an official guidebook:** U.S. Department of Health, Education, and Welfare, *The Institutional Guide to DHEW Policy on Protection of Human Subjects.* DHEW Publication No. (NIH) 72-102 (Washington, DC: Government Printing Office, 1971). There are several unofficial guidebooks today, but no official one.

23      **was triggered by a question:** House of Representatives, Subcommittee on Human Resources, Committee on Government Reform and Oversight (1998), *Institutional Review Boards: A System in Jeopardy,* 105th Congress, Second Session, Serial No. 105-166.

24      **took the hammer off the shelf:** Jeffrey Brainard and D. W. Miller,"Spate of Suspensions of Academic Research Spurs Questions about Federal Strategy," *The Chronicle of Higher Education* 46, no. 22 (Feb 4, 2000): A29.

24      **a dozen other institutions:** Brainard and Miller, "Spate of Suspensions."

24      **They would rather be foolish:** Caroline H. Bledsoe, Bruce Sherin, Adam G. Galinsky, Nathalia M. Headley, Carol A. Heimer, Erik Kjeldgaard, James Lindgren, Jon D. Miller, Michael E. Roloff, and David H. Uttal, "Regulating Creativity: Research and Survival in the IRB Iron Cage," *Northwestern University Law Review* 101, no. 2 (Feb 1, 2007): 593-641.

26      **people with a conflict of interest:** Dennis John Mazur, *Evaluating the Science and Ethics of Research on Humans: A Guide for IRB Members* (Baltimore: Johns Hopkins University Press, 2007), p. 3.

26      **to see scientists as corruptible:** This attitude—that IRBs defend morality and scientists lack it—is on conspicuous display in the online training materials that almost every research university requires its scientists to complete in order to conduct human subjects research.

26      **he wrote an article:** Coe, "Costs and Benefits."

26      **Not all scientists oppose the system:** Simon N. Whitney, Kirsten Alcser, Carl Schneider, Laurence B. McCullough, Amy L. McGuire, and Robert J. Volk, "Principal Investigator Views of the IRB System,"

*International Journal of Medical Sciences* 5, no. 2 (Apr 2, 2008): 68-72.

26    **will make the survey appear comical:** I am not making this up. I have gotten surveys with this warning myself.

26    **might move people to suicide:** Dennis Mazur, in his IRB manual, advises that people answering a questionnaire on post-traumatic stress disorder may have reactions "that could include suicidal ideation or even suicide attempt." Mazur, *Evaluating the Science*, p. 19. I have asked Mazur for evidence for this assertion and he could provide none; emails on file with author.

27    **useful and risk-free study:** Ethicists, and regulators, are ready to point out that no experiment is free of risk. Coe could take the anonymous urine samples in his lab, analyze their DNA, match them to genetic databases, and identify the subjects. If his database is hacked, his subjects' names could be released, perhaps on the internet. An insurance company could find the names, learn that some of them have kidney stones, and refuse to sell them health insurance. Harms like this are too far-fetched to discuss in this introductory chapter, but they are too often raised to ignore. We will review them later in the book.

27    **the ethicists die of diseases:** Whitney, "Principal Investigator Views."

27    **changes in the regulations:** Department of Homeland Security and 15 other departments and agencies, *Final Rule, 82 Fed Reg 7149-7273* (January 19, 2017); Department of Homeland Security and 15 other departments and agencies, *Federal Policy for the Protection of Human Subjects: Delay of the Revisions to the Federal Policy for the Protection of Human Subjects, 83 Fed Reg 2885-94* (January 22, 2018).

27    **urine samples from normal people:** Email from Fredric Coe, March 1, 2020.

28    **he calls it a tyranny:** Coe, "Costs and Benefits."

29    **modern methods of risk management:** For a general introduction to risk management theory and practice, see Peter L. Bernstein, *Against the Gods: The Remarkable Story of Risk* (New York: John Wiley & Sons, 1996) and Paul Hopkin, *Fundamentals of Risk Management,* 4th ed (New York, NY: Kogan Page, 2017).

29     **behind contemporary norms:** Simon N. Whitney, "A Fern in Amber: Risk Management in Research with Humans," *Risk Management* 17, no. 4 (2015): 226-239; Simon N. Whitney, "Institutional Review Boards: A Flawed System of Risk Management," *Research Ethics* 12, no. 4 (2016): 182-200.

## 2. The Birth of Research Oversight

31     **Patient R.M.:** Charles H. Burnett, Esther L. Bloomberg, Gerald Shortz, David W. Compton, and Henry K. Beecher, "A Comparison of the Effects of Ether and Cyclopropane Anesthesia on the Renal Function of Man," *Journal of Pharmacology and Experimental Therapeutics* 96, no. 4: 380-387. I am grateful to David Jones, Christine Grady, and Susan Lederer for drawing attention to this experiment; see David S. Jones, Christine Grady, and Susan E. Lederer, "'Ethics and Clinical Research'— the 50th Anniversary of Beecher's Bombshell," *The New England Journal of Medicine* 374, no. 24 (2016): 2393-2398.

32     **I am Doctor Nathan:** John Lancaster and James P. Rathwell, "The Moralist," *The Medium*, July 14, 2016, accessed October 16, 2022, https://medium.com/thebigroundtable/the-moralist-ad8159ebe6be.

33     **fortunately temporary effects:** Keith Sykes, "Fifty Years On," *History of Anaesthesia Society Proceedings* 34, (2004): 9-15.

33     **Beecher eventually came to feel:** Jones, Grady, and Lederer, "Ethics and Clinical Research."

33     **James Shannon:** Thomas J. Kennedy, Jr., "James Augustine Shannon," in *Biographical Memoirs* (Washington, DC: National Academies Press, 1994).

34     **measured in human lives:** Lancaster, "The Moralist."

35     **the Clinical Research Committee:** Mark S. Frankel, "The Public Health Service Guidelines Governing Research Involving Human Subjects: An Analysis of the Policy-Making Process," PhD dissertation, George Washington University, 1972; Laura Jeanine Morris Stark, *Behind Closed Doors: IRBs and the Making of Ethical Research* (Chicago; London: The University of Chicago Press, 2012).

35     **moved smartly ahead:** As historian and ethicist Susan Lederer writes, "The development of penicillin, cortisone, gamma globulin,

blood substitutes, and other drugs and techniques impressed both Congress and the American public. ... In 1946 the NIH received approximately $700,000 from the federal government. By 1955 the NIH appropriation exceeded $36 million; in 1970, NIH received nearly $1.5 billion from the federal government and administered over 11,000 grants." Susan E. Lederer, *Subjected to Science: Human Experimentation in America Before the Second World War* (Baltimore: Johns Hopkins University Press, 1995), 140.

35    **NIH officials regularly advised:** Frankel, "Public Health Service Guidelines."

36    **killed an experimental subject:** Stark, *Behind Closed Doors*.

36    **no specific therapy was available:** Leonard D. Heaton, John B. Coates Jr., Ebbe C. Hoff and Phebe M. Hoff, "Preventive Medicine in World War II. Volume 5. Communicable Diseases Transmitted through Contact Or by Unknown Means," Office of the Surgeon General (Army), 1960.

36    **killed thousands more:** Herrman L. Blumgart, "The Medical Framework for Viewing the Problem of Human Experimentation," *Daedalus* (1969): 248-274.

38    **the problem disappeared:** Richard B. Capps, Alfred M. Bennett, and Joseph Stokes, "Endemic Infectious Hepatitis in an Infants' Orphanage: I. Epidemiologic Studies in Student Nurses," *AMA Archives of Internal Medicine* 89, no. 1 (1952): 6-23; Joseph Stokes, Jr., "Epidemiology of Viral Hepatitis A," *American Journal of Public Health and the Nation's Health* 43, no. 9 (Sep, 1953): 1097-1100; Joseph Stokes Jr., J. Edward Berk, Leonard L. Malamut, Miles E. Drake, Jeremiah A. Barondess, Winslow J. Bashe, Irving J. Wolman, John D. Farquhar, B. Bevan, and RJ Drummond, "The Carrier State in Viral Hepatitis," *Journal of the American Medical Association* 154, no. 13 (1954): 1059-1065.

38    **mild symptoms:** Paul A. Offit, *Vaccinated: One Man's Quest to Defeat the World's Deadliest Diseases* (New York: Harper Collins, 2008).

39    **between days 26 and 66:** Saul Krugman, Robert Ward, Joan P. Giles, Oscar Bodansky, and A. Milton Jacobs, "Infectious Hepatitis: Detection of Virus during the Incubation Period and in Clinically

Inapparent Infection," *New England Journal of Medicine* 261, no. 15 (1959): 729-734.

40    **open windows to eternity:** J.S. Gravenstein, "Henry K. Beecher: The Introduction of Anesthesia into the University," *Anesthesiology* 88, no. 1 (1998): 245-253.

41    **to avoid at all costs:** Henry K. Beecher, "Ethics and Clinical Research," *The New England Journal of Medicine* 274, no. 24 (June 16, 1966): 1354-1360.

41    **materials for a major article:** This was to become Henry K. Beecher, "Experimentation in Man," *Journal of the American Medical Association* 169, no. 5 (1959): 461-478.

42    **the Nuremberg Code:** http://www.hhs.gov/ohrp/archive/-nurcode.html.

43    **more harm than good:** Beecher, "Experimentation in Man."

44    **happily endorsed:** David J. Rothman, *Strangers at the Bedside: A History of how Law and Bioethics Transformed Medical Decision Making* (New York, NY: Basic Books, 1991).

44    **doing anything about it:** Irving Ladimer and Roger W. Newman, *Clinical Investigation in Medicine: Legal, Ethical, and Moral Aspects: an Anthology and Bibliography* (Boston: Law-Medicine Research Institute, Boston University, 1963).

45    **guinea pigs, rats, and mice:** G. McConnell, "The Transplantation of Human Carcinomatous Material into Lower Animals," *The Journal of Experimental Medicine* 10, no. 1 (Jan 1, 1908): 36-44.

46    **the swelling soon disappeared:** Nicholas Senn, "A Plea For the International Study of Carcinoma," *Journal of the American Medical Association* XLVI, no. 17 (1906): 1254-1258.

47    **Henrietta Lacks's biopsy:** Rebecca Skloot, *The Immortal Life of Henrietta Lacks* (New York: Crown Publishers, 2010).

47    **would be made no shorter:** Jay Katz, Alexander Morgan Capron, and Eleanor Swift Glass, *Experimentation with Human Beings: the Authority of the Investigator, Subject, Professions, and State in the Human Experimentation Process* (New York: Russell Sage Foundation, 1972), 36-37.

48    **diameter of about an inch:** Chester M. Southam, Alice E. Moore, and Cornelius P. Rhoads, "Homotransplantation of Human Cell Lines," *Science* 125, no. 3239 (1957): 158-160.

**49**     **payment or early parole:** Ruth Brecher and Edward Brecher, "They Volunteered for Cancer," *Reader's Digest,* April, 1958, 62-66.

**50**     **the Jewish Chronic Disease Hospital:** Katz, Capron, and Glass, *Experimentation with Human Beings,* 9-65.

**51**     **advanced Parkinson's dementia:** Katz, Capron, and Glass, *Experimentation with Human Beings,* 13 and 18.

**51**     ***Times* front-page headline:** James McCaffrey, "Hospital Accused on Cancer Study," *New York Times* January 21, (1964).

**51**     **Senator Jacob Javits:** Laura Jeanine Morris Stark, *Behind Closed Doors: IRBs and the Making of Ethical Research* (Chicago; London: The University of Chicago Press, 2012), 149-151.

**52**     **no greater sacrifice:** Rothman, *Strangers at the Bedside,* 36-49.

**52**     **and the mentally ill:** Ruth R. Faden, Tom L. Beauchamp, and Nancy M. P. King, *A History and Theory of Informed Consent* (New York: Oxford University Press, 1986), 87.

**52**     **a feather in a high wind:** Vincent J. Kopp, "Henry Knowles Beecher and the Development of Informed Consent in Anesthesia Research," *Anesthesiology: The Journal of the American Society of Anesthesiologists* 90, no. 6 (1999): 1756-1765.

**53**     **admitted frankly:** Lancaster, "The Moralist."

**53**     **shaking the medical establishment:** Jones, Grady, and Lederer, "Ethics and Clinical Research."

**54**     **and he sued:** Katz, Capron, and Glass, *Experimentation with Human Beings,* 40-41.

**54**     **public relations exposure:** Mark S. Frankel, "The Public Health Service Guidelines Governing Research Involving Human Subjects: An Analysis of the Policy-Making Process," PhD dissertation, George Washington University, 1972, 23-24.

**54**     **gaining useful information:** Frankel, "Public Health Service Guidelines," 15.

**54**     **in defense of subjects' rights:** Frankel, "Public Health Service Guidelines."

**54**     **Shannon and other top officials:** Frankel, "Public Health Service Guidelines"; Rothman, *Strangers at the Bedside*; Laura Jeanine Morris Stark, *Behind Closed Doors: IRBs and the Making of Ethical Research* (Chicago; London: The University of Chicago Press, 2012).

55      **Public Health Service Policy #129:** Office of the Surgeon General, US Public Health Service. "Surgeon General's Directives on Human Experimentation." *American Psychologist* 22, (1967): 350-355.

56      **a "dramatic" decline:** William J. Curran, "Governmental Regulation of the use of Human Subjects in Medical Research: The Approach of Two Federal Agencies," *Daedalus* 98, no. 2 (1969): 542-594.

56      **the development of an ethical code:** Frankel, "Public Health Service Guidelines."

57      **he is best remembered:** Henry K. Beecher, "Ethics and Clinical Research," *The New England Journal of Medicine* 274, no. 24 (June 16, 1966): 1354-1360.

58      **widespread media attention:** Lara Freidenfelds, "Recruiting Allies for Reform: Henry Knowles Beecher's 'Ethics and Clinical Research'," *International Anesthesiology Clinics* 45, no. 4 (2007): 79-103.

59      **your honesty and courage:** Freidenfelds, "Recruiting Allies for Reform."

59      **quotes out of context:** James L. Scott, Gerald A. Belkin, Sydney M. Finegold, and John S. Lawrence, "Human Experimentation (Letter)," *New England Journal of Medicine* 275, no. 14 (1966): 790-791.

59      **with forcing Shannon's hand:** Rothman, *Strangers at the Bedside*; Kopp, "Henry Knowles Beecher"; Freidenfelds, "Recruiting Allies for Reform"; M. Best and D. Neuhauser, "Henry K Beecher: Pain, Belief and Truth at the Bedside. the Powerful Placebo, Ethical Research and Anaesthesia Safety," *Quality & Safety in Health Care* 19, no. 5 (Oct, 2010): 466-468; Robert L. Klitzman, *The Ethics Police?: The Struggle to make Human Research Safe* (Oxford; New York: Oxford University Press, 2015).

59      **Shannon was aware:** Frankel, "Public Health Service Guidelines." Thanks to Zachary Schrag for drawing this to my attention.

59      **finesse some awkward problems:** His research with R.M. was not the only problem Beecher confronted. He had conducted research for the CIA as part of the MKULTRA project and had studied the effects of LSD with pharmacologist Louis Lasagna. This research might be seen as a more obvious example of Beecher's moral failings,

but our inquiry focuses on the ethics of ordinary medical research, not on what research should be conducted under a perceived threat to national security. For more information, see Select Committee on Intelligence, United States Senate, Project MKULTRA, the CIA's Program of Research in Behavioral Modification, Ninety-fifth Congress, First Session, August 3, 1977; Lancaster, "The Moralist"; George A. Mashour, "Altered States: LSD and the Anesthesia Laboratory of Henry Knowles Beecher," *Environment* 69, (2009); and George A. Mashour, "From LSD to the IRB: Henry Beecher's Psychedelic Research and the Foundation of Clinical Ethics," *International Anesthesiology Clinics* 45, no. 4 (2007): 105-111.

59    **the experiment is without risk:** Beecher, "Experimentation in Man."

59    **seventy-four minutes longer:** Charles H. Burnett, Esther L. Bloomberg, Gerald Shortz, David W. Compton, and Henry K. Beecher, "A Comparison of the Effects of Ether and Cyclopropane Anesthesia on the Renal Function of Man," *Journal of Pharmacology and Experimental Therapeutics* 96, no. 4: 380-387.

60    **his own problematic research:** Beecher, "Ethics and Clinical Research."

60    **the CRC's review process:** Beecher, Henry K. "Experimentation in Man." *Journal of the American Medical Association* 169, no. 5 (1959): 461-478.

60    **by four separate committees:** Saul Krugman, "The Willowbrook Hepatitis Studies Revisited: Ethical Aspects," *Reviews of Infectious Diseases* 8, no. 1 (1986): 157-162.

61    **the AAAS held conferences:** The proceedings were published in *Daedalus.* See "Ethical Aspects of Experimentation with Human Subjects," *Daedalus* 98, no. 2 (Spring 1969).

61    **how risky the experiment was:** Beecher, "Ethics and Clinical Research."

61    **the Nazi shudders:** Jones, Grady, and Lederer, "Ethics and Clinical Research," quoting Beecher papers in the Harvard Medical Library, Francis A. Countway Library of Medicine.

61    **Art Caplan:** Offit, *Vaccinated.*

61    **unethical medical experiments:** Offit, *Vaccinated.*

62    **the study was unobjectionable:** Louis Lasagna, "Special Subjects in Human Experimentation," *Daedalus* 98, no. 2 (1969): 449-462.

62    **an inconsequential injury?** Walter Goodman, "Doctors must Experiment on Humans but What are the Patient's Rights?" *New York Times Magazine* July 2, 1967.

62    **sometimes-fatal infections:** CDC, "Global Immunization," updated July 27, 2022, accessed 7/27/22 https://www.cdc.gov/-globalhealth/immunization/diseases/hepatitis-b/data/fast-facts.html.

63    **children, born and unborn:** Krugman, "The Willowbrook Hepatitis Studies Revisited."

64    **it was time to flee:** David J. Levy, *Hans Jonas: The Integrity of Thinking*, Vol. 1 (Columbia, MO: University of Missouri Press, 2002).

65    **progress is an optional goal:** Hans Jonas, "Philosophical Reflections on Experimenting with Human Subjects," *Daedalus* 98, no. 2 (1969): 219-247.

65    **required no consent at all:** Beecher, "Experimentation in Man."

66    **situations of social life:** Jonas, "Philosophical Reflections."

67    **full-time IRB staff:** Sarah Babb reviews this process in detail in her insightful book, *Regulating Human Research: IRBs from Peer Review to Compliance Bureaucracy* (Stanford, California: Stanford University Press, 2020).

**3. Congress Steps In**

69    **a fist-sized bulge:** Peter Buxton, "Testimony by Peter Buxton from the United States Senate Hearings on Human Experimentation," in *Tuskegee's Truths: Rethinking the Tuskegee Syphilis Study*, ed. Susan Reverby (Chapel Hill and London: University of North Carolina Press, 2000), pp. 150-156. His name is usually spelled Buxtun, but in the senate hearings it is reported as Buxton.

69    **a libertarian Republican**: Susan Reverby, *Examining Tuskegee: The Infamous Syphilis Study and its Legacy* (Chapel Hill: University of North Carolina Press, 2009).

69    **when he was an infant:** Lawrence Bush, "July 29: The Tuskegee Syphilis Experiment," *Jewish Currents,* July 28, 2015.

**70**    **they were being denied care:** The story of what is properly called the United States Public Health Service Syphilis Study has been told many times. Historian James Jones published the first full history, historian Susan Reverby wrote two volumes that are essential to a full understanding, and preacher and lawyer Fred Gray has told the story from the inside, beginning with the desperate poverty of Black sharecroppers in 1930s Alabama. James H. Jones, *Bad Blood: The Tuskegee Syphilis Experiment* (New York and London: Free Press, 1981); Reverby, *Tuskegee's Truths*; Reverby, *Examining Tuskegee*; Fred Gray, *The Tuskegee Syphilis Study: The Real Story and Beyond* (Montgomery, AL: NewSouth Books, 1998).

**70**    **political dynamite:** Reverby, *Examining Tuskegee*, 79.

**70**    **until the last subject died:** Reverby, *Tuskegee's Truths*, 471.

**70**    **America's front pages:** Jones, *Bad Blood*, 204.

**71**    **arsenic and bismuth:** Gray, *The Tuskegee Syphilis Study*.

**71**    **bore infected children:** M. Joycelyn Elders, Rueben C. Warren, Vivian W. Pinn, James H. Jones, Susan M. Reverby, David Satcher, Mary E. Northridge, Ronald Braithwaite, Mario DeLaRosa, and Luther S. Williams, *The Search for the Legacy of the USPHS Syphilis Study at Tuskegee: Reflective Essays Based upon Findings from the Tuskegee Legacy Project*, ed. by Ralph V. Katz and Rueben C. Warren (Lanham, Maryland: Lexington Books, 2011).

**71**    **rankles Tuskegee University:** Tuskegee University, "About the USPHS Syphilis Study," n.d, accessed November 3, 2022, https://-www.tuskegee.edu/about-us/centers-of-excellence/-bioethics-center/about-the-usphs-syphilis-study.

**72**    **presumption of infallibility:** U.S. Department of Health, Education, and Welfare, *The Institutional Guide to DHEW Policy on Protection of Human Subjects (the Yellow Book)*, DHEW Publication No. (NIH) 72-102 (Washington, DC: Government Printing Office, 1971), p. 9, section C.

**72**    **one IRB manual explains:** There are several IRB manuals, all unofficial.

**72**    **may have different agendas:** Ernest D. Prentice, Sally L. Mann, and Bruce G. Gordon, "Administrative Reporting Structure for the Institutional Review Board," in *Institutional Review Board: Management and Function*, 2nd ed, ed. by Elizabeth A. Bankert and

Robert J. Amdur (Burlington, MA: Jones & Bartlett Learning, 2006), 31-32.

73    **jealousy and academic infighting:** Stephen J. Ceci, Douglas Peters, and Jonathan Plotkin, "Human Subjects Review, Personal Values, and the Regulation of Social Science Research," *The American Psychologist* 40, no. 9 (1985): 994-1002; Daniel K. Nelson, "Conflict of Interest: Institutional Review Boards," in *Institutional Review Board: Management and Function*, 2nd ed, ed. by Elizabeth A. Bankert and Robert J. Amdur (Burlington, MA: Jones & Bartlett Learning, 2006), 208-212; Laura Jeanine Morris Stark, *Behind Closed Doors: IRBs and the Making of Ethical Research* (Chicago; London: The University of Chicago Press, 2012); Robert L. Klitzman, *The Ethics Police?: The Struggle to make Human Research Safe* (Oxford; New York: Oxford University Press, 2015).

73    **National Research Act:** National Research Service Award Act of 1974, Pub. L. No. 93-348, 88 Stat. 342.

73    **will require Congressional action:** The statute did not require that the regulations be so restrictive, so a measure of reform could be achieved without Congressional action. But the statute, the regulations, and the spirit of the enterprise are so linked, and so unbalanced, that only by fixing all three can we achieve true reform. Still IRBs could do better, even in the present system, as I explain in my own manual—Simon N. Whitney, *Balanced Ethics Review: A Guide for Institutional Review Board Members* (Cham, Switzerland: Springer, 2016).

74    **Congressional hearings:** House of Representatives, Subcommittee on Human Resources, Committee on Government Reform and Oversight (1998), *Institutional Review Boards: A System in Jeopardy,* 105th Congress, Second Session, Serial No. 105-166.

75    **I don't like the word 'reform':** House of Representatives, *Institutional Review Boards.*

75    **oversight activities are working:** General Accounting Office, *Scientific Research: Continued Vigilance Critical to Protecting Human Subjects: Report to the Ranking Minority Member, Committee on Governmental Affairs, U.S. Senate* (District of Columbia: General Accounting Office, 1996), GAO/HEHS-96-72.

76 **shielded from unnecessary harm:** You may wonder, what is "necessary harm"? For healthy subjects who are enrolled in tests of new treatments, necessary harm is limited to side effects that are mild or moderate. For the sick, diseases and their treatments both bring harms, whether the patient is treated in a routine clinic or as part of an experimental trial, and this influences the harms that can be attributed to the experiment itself.

77 **the third priority is the researcher:** Zachary M. Schrag, *Ethical Imperialism: Institutional Review Boards and the Social Sciences, 1965-2009* (Baltimore: Johns Hopkins University Press, 2010), 134.

77 **destroy this one and start over:** House of Representatives, *Institutional Review Boards.*

77 **pressure motivated Ellis:** Ellis was also influenced by pressure from advocacy groups and, later, from a national commission on bioethics. But it was Congressional pressure, and presumably the threat of action against his own agency, that led him to impose the death penalty so liberally. Eliot Marshall, "Shutdown of Research at Duke Sends a Message," *Science* 284, no. 5418 (1999): 1246.

77 **a dozen institutions:** Sydney Halpern, "Hybrid Design, Systemic Rigidity: Institutional Dynamics in Human Research Oversight," *Regulation & Governance* 2, no. 1 (Mar, 2008): 85-102.

78 **not particularly hazardous:** Robert J. Levine, *Ethics and Regulation of Clinical Research*, 2nd ed. (Baltimore: Urban & Schwarzenberg, 1986), 40; Levine testimony in House of Representatives, *Institutional Review Boards.*

78 **never previously been observed:** Steven E. Raper, Narendra Chirmule, Frank S. Lee, Nelson A. Wivel, Adam Bagg, Guang-ping Gao, James M. Wilson, and Mark L. Batshaw, "Fatal Systemic Inflammatory Response Syndrome in a Ornithine Transcarbamylase Deficient Patient Following Adenoviral Gene Transfer," *Molecular Genetics and Metabolism* 80, no. 1-2 (2003): 148-158.

78 **reached for any handy excuse:** For a look at the death penalty era from the point of view of the regulators, who considered their actions fully justified, see Sarah L. Babb, *Regulating Human Research: IRBs from Peer Review to Compliance Bureaucracy* (Stanford, California: Stanford University Press, 2020).

78    **to document their compliance:** Joan E. Sieber, Stuart Plattner, and Rubin Philip, "How (not) to regulate social and behavioral research," *Professional Ethics Report* 15(2), Spring 2002.

78    **Virginia Commonwealth University:** Jay Mathews, "Father's Complaints Shut Down Research," *Washington Post*, January 12, 2000.

78    **before they could resume:** Jeffrey Brainard and D. W. Miller, "Spate of Suspensions of Academic Research Spurs Questions about Federal Strategy," *The Chronicle of Higher Education* 46, no. 22 (Feb 4, 2000): A29: Rick Weiss, "US Halts Human Research at Duke," *Washington Post* May 12, 1999, p. A1.

79    **one-half plus one:** Norman Fost and Robert J. Levine, "The Dysregulation of Human Subjects Research," *JAMA: The Journal of the American Medical Association* 298, no. 18 (Nov 14, 2007).

79    **months, if not years, of work:** J. Michael Oakes, "Risks and Wrongs in Social Science Research," *Evaluation Review* 26, no. 5 (Oct, 2002): 443-479.

79    **research funds ... at stake:** Brainard and Miller, "Spate of Suspensions."

79    **suspension of federal funding:** Caroline H. Bledsoe, Bruce Sherin, Adam G. Galinsky, Nathalia M. Headley, Carol A. Heimer, Erik Kjeldgaard, James Lindgren, Jon D. Miller, Michael E. Roloff, and David H. Uttal, "Regulating Creativity: Research and Survival in the IRB Iron Cage," *Northwestern University Law Review* 101, no. 2 (Feb 1, 2007): 593-641.

79    **discredit the IRB:** C. Kristina Gunsalus, Edward M. Bruner, Nicholas C. Burbules, Leon Dash, Matthew Finkin, Joseph P. Goldberg, William T. Greenough, et al, "The Illinois White Paper: Improving the System for Protecting Human Subjects: Counteracting IRB 'Mission Creep'," *Qualitative Inquiry* 13, no. 5 (Jul, 2007): 617-649.

80    **transformed IRBs nationwide:** Sarah L. Babb, *Regulating Human Research: IRBs from Peer Review to Compliance Bureaucracy* (Stanford, California: Stanford University Press, 2020). The transformation of IRBs has been described by many other scholars, including Bledsoe et al. ("Regulating Creativity") and Gunsalus et al. ("Illinois White Paper").

81    **the Northwestern IRB:** Todd J. Zywicki, "Institutional Review Boards as Academic Bureaucracies: An Economic and Experiential Analysis," *Northwestern University Law Review* 101, no. 2 (Feb 1, 2007): 861-865.

81    **avoid the enormous risk:** Bledsoe et al., "Regulating Creativity."

81    **multiple levels of obstruction:** Babb, *Regulating Human Research*, 46.

82    **it scared me to death:** Norman M. Goldfarb, "Greg Koski on Human Subjects Protection," *Journal of Clinical Research and Best Practices* 3, no. 9 (2007): 1-6.

82    **should be avoided:** Greg Koski, "Beyond Compliance...is it Too Much to Ask?" *IRB: Ethics & Human Research* 25, no. 5 (2003): 5-6.

## 4. Risks and Benefits, Real and Imaginary

84    ***Dirt is Good:*** Jack Gilbert, Rob Knight, and Sandra Blakeslee, *Dirt is Good: The Advantage of Germs for Your Child's Developing Immune System* (New York, NY: St. Martin's Press, 2017).

85    **espionage or time travel:** Rob Knight, personal communications, 2010, 2013, and 2021.

85    **practicing doctors and active scientists:** University of Colorado, "IRB Rosters," August 15, 2022, accessed on November 4, 2022 at https://research.cuanschutz.edu/comirb/home/irb_members/irb-rosters#ft-current-roster-0.

85    **come from IRB staff:** Sarah L. Babb, *Regulating Human Research: IRBs from Peer Review to Compliance Bureaucracy* (Stanford, California: Stanford University Press, 2020).

86    **change an IRB's decision:** Laura Jeanine Morris Stark, *Behind Closed Doors: IRBs and the Making of Ethical Research* (Chicago; London: The University of Chicago Press, 2012), 30.

87    **slight, minute, and atomic in scale:** Fredric L. Coe, "Costs and Benefits of a Well-Intended Parasite: A Witness and Reporter on the IRB Phenomenon," *Northwestern University Law Review* 101, (2007): 723-733.

88    **subjects in genetic studies:** Ethicists Amy McGuire and Mats Hansson, law professors Henry Greely and Barbara Evans, and their collaborators are among those who have explored the evolving field of

genetics and genomics and its implications for cutting-edge research. See, for instance, Amy L. McGuire and Laura M. Beskow, "Informed Consent in Genomics and Genetic Research," *Annual Review of Genomics and Human Genetics* 11, (2010): 361; Mats Hansson, "Genomics and society—Ethical, Legal, Cultural and Socioeconomic Implication," *European Journal of Human Genetics* 24, no. 12 (2016): 1830; Jennifer Kulynych and Henry T. Greely, "Clinical Genomics, Big Data, and Electronic Medical Records: Reconciling Patient Rights with Research when Privacy and Science Collide," *Journal of Law and the Biosciences* 4, no. 1 (2017): 94-132; Henry T. Greely, "Legal, Ethical, and Social Issues in Human Genome Research," *Annual Review of Anthropology* (1998): 473-502; Barbara J. Evans, "The First Amendment Right to Speak about the Human Genome," *University of Pennsylvania Journal of Constitutional Law* 16, no. 3 (2014): 549; Barbara J. Evans, "The Genetic Information Nondiscrimination Act at Age 10: GINA's Controversial Assertion that Data Transparency Protects Privacy and Civil Rights," *William and Mary Law Review* 60, no. 6 (2019): 2017; Susan M. Wolf, Frances P. Lawrenz, Charles A. Nelson, Jeffrey P. Kahn, Mildred K. Cho, Ellen Wright Clayton, Joel G. Fletcher, Michael K. Georgieff, Dale Hammerschmidt, and Kathy Hudson, "Managing Incidental Findings in Human Subjects Research: Analysis and Recommendations," *Journal of Law, Medicine & Ethics* 36, no. 2 (2008): 219-248.

88    **Get out of the way:** Steven Pinker, "The Moral Imperative for Bioethics," *Boston Globe* August 1, 2015, accessed August 19, 2019, https://www.bostonglobe.com/opinion/2015/07/31/the-moral-imperative-for-bioethics/JmEkoyzlTAu9oQV76JrK9N/story.html.

89    **illustrate risks to subjects:** Joan E. Sieber, *Planning Ethically Responsible Research: A Guide for Students and Internal Review Boards* (Newbury Park; London; New Delhi: SAGE Publications, 1992), 80.

90    **the project should be approved:** Maureen H. Fitzgerald and Paul A. Phillips, "Centralized and Non-Centralized Ethics Review: A Five Nation Study," *Accountability in Research* 13, no. 1 (2006): 47-74.

90    **subjects could be warned:** Jonathan Baron, "Some Fallacies of Human-Subjects Protection, and some Solutions," *Cortex; a Journal*

*Devoted to the Study of the Nervous System and Behavior* 65, (2014): 246-254.

91   **particularly risk-averse:** Pavel D. Atanasov, "Risk Preferences in Choices for Self and Others: Meta Analysis and Research Directions," *Social Science Research Network* (posted 9/25/2010, last revised 6/17/2015), doi:10.2139/ssrn.1682569.

91   **vulnerable, risk-averse participants:** Michelle N. Meyer, "Three Challenges for Risk-Based (Research) Regulation: Heterogeneity among Regulated Activities, Regulator Bias, and Stakeholder Heterogeneity," in I. Glenn Cohen and Holly Fernandez Lynch, eds, *Human Subjects Research Regulation: Perspectives on the Future* (Cambridge, MA: MIT Press, 2014).

92   **balanced or objective position:** Paul Ramsey, *The Patient as Person: Explorations in Medical Ethics* (New Haven: Yale University Press, 1970), 10.

92   **duped into participation:** The federal investigation into the gene transfer experiment that resulted in the death of Jesse Gelsinger in 1999 is a good example. See FDA, Warning Letter to Mark Batshaw, November 30, 2000, signed by Steven Masiello, Director, Office of Compliance and Biologics Quality, on file with author and FDA, Notice of Opportunity for Hearing to James Wilson, February 8, 2002, signed by Dennis Baker, Associate Commissioner for Regulatory Affairs, on file with author.

92   **their values and preferences:** Jonathan Baron, *Against Bioethics* (Cambridge, MA: MIT Press, 2006), 118.

93   **we do know something:** For a thoughtful exploration of the issues involved, see Franklin G. Miller and Howard Brody, "What Makes Placebo-Controlled Trials Unethical?" *The American Journal of Bioethics* 2, no. 2 (2002): 3-9.

93   **wronged even when not harmed:** Hans Jonas, "Philosophical Reflections on Experimenting with Human Subjects," *Daedalus* 98, no. 2 (1969): 219-247.

93   **being treated by their own doctor:** Charles Fried, *Medical Experimentation: Personal Integrity and Social Policy* (Amsterdam; New York: North-Holland Pub. Co.; American Elsevier, 1974).

93    **These are the tragic choices:** Jerry Menikoff and Edward P. Richards, *What the Doctor Didn't Say: The Hidden Truth about Medical Research* (Oxford: Oxford University Press, 2006), 18-19.

94    **a small amount of inconvenience:** Menikoff and Richards, *What the Doctor Didn't Say,* 17-18.

94    **reviewed cancer studies:** J.M. Peppercorn, J. C. Weeks, E. F. Cook, and S. Joffe, "Comparison of Outcomes in Cancer Patients Treated within and Outside Clinical Trials: Conceptual Framework and Structured Review," *Lancet* 363, no. 9405 (Jan 24, 2004): 263-270.

94    **either harm or benefit:** Gunn Elisabeth Vist, Dianne Bryant, Lyndsay Somerville, Trevor Birminghem, and Andrew D. Oxman, "Outcomes of Patients Who Participate in Randomized Controlled Trials Compared to Similar Patients Receiving Similar Interventions Who do Not Participate," *Cochrane Library* 2010, no. 1 (Jul 16, 2008): MR000009, doi:10.1002/14651858.MR000009.pub4.

95    **to lead to better outcomes:** John D. Lantos, "The 'Inclusion Benefit' in Clinical Trials," *The Journal of Pediatrics* 134, no. 2 (Feb, 1999): 130-131.

95    **well-supervised research study:** Patricia C. El-Hinnawy, "Interview with Norman Fost" in *OHRP: Oral History of the Belmont Report and the National Commission for the Protection of Human Subjects of Biomedical and Behavioral Research* (Belmont Oral History Project, 2004), accessed on December 26, 2022 at https://www.hhs.gov/ohrp/education-and-outreach/luminaries-lecture-series/belmont-report-25th-anniversary-interview-nfost/index.html.

95    **as many as 28,000:** Sean Berenholtz, Peter Pronovost, Pamela Lipsett, Deborah Hobson, Karen Earsing, Jason Farley, Shelley Milanovich, et al., "Eliminating Catheter-Related Bloodstream Infections in the Intensive Care Unit," *Critical Care Medicine* 32, no. 10 (Oct, 2004): 2014-2020; Peter Pronovost, Dale Needham, Sean Berenholtz, David Sinopoli, Haitao Chu, Sara Cosgrove, Bryan Sexton, Robert Hyzy, Robert Welsh, and Gary Roth, "An Intervention to Decrease Catheter-Related Bloodstream Infections in the ICU," *New England Journal of Medicine* 355, no. 26 (2006): 2725-2732.

96    **forgot one or more steps:** Berenholtz et al., "Eliminating Catheter-Related Bloodstream Infections."

97    **saved almost $2 million:** Berenholtz et al., "Eliminating Catheter-Related Bloodstream Infections."

97    **a corrective action plan:** Kristina C. Borror, Director, Division of Compliance Oversight, OHRP, *Letter to Daniel E. Ford, Vice Dean for Clinical Investigation, Johns Hopkins, July 19, 2007,* re ICU infection study.

97    **Congress will have to:** Atul Gawande, "A Lifesaving Checklist" (Op-Ed), *New York Times,* December 30, 2007. There was plenty of criticism from academic ethicists as well. Ethicist Mary Faith Marshall lacerated OHRP's intervention in a talk at the 2007 meeting of PRIM&R, the IRB professional and advocacy organization.

98    **but it's possible:** Bridget M. Kuehn, "DHHS Halts Quality Improvement Study," *JAMA: The Journal of the American Medical Association* 299, no. 9 (Mar 5, 2008): 1005-1006.

98    **people who will die unnecessarily:** Ruth R. Faden, Nancy E. Kass, Steven N. Goodman, Peter Pronovost, Sean Tunis, and Tom L. Beauchamp, "An Ethics Framework for a Learning Health Care System: A Departure from Traditional Research Ethics and Clinical Ethics," *The Hastings Center Report* 43, no. s1 (Jan, 2013): S16-S27.

## 5. Complicated Consent

99    **puts you in a sling:** Your broken clavicle cannot be put in a cast, so a sling, or sometimes a special splint, are appropriate immediate care.

100   **to join the trial:** Smekal and colleagues have done just such a trial (overall, patients who had surgery did better). See V. Smekal, A. Irenberger, P. Struve, M. Wambacher, D. Krappinger, and F. S. Kralinger. "Elastic Stable Intramedullary Nailing Versus Nonoperative Treatment of Displaced Midshaft Clavicular Fractures—A Randomized, Controlled, Clinical Trial." *Journal of Orthopaedic Trauma* 23, no. 2 (Feb, 2009): 106-112.

100   **not that stuff on the paper:** Fredric L. Coe, "Costs and Benefits of a Well-Intended Parasite: A Witness and Reporter on the IRB Phenomenon," *Northwestern University Law Review* 101, (2007): 723-733.

100    **a decision already made:** Carol A. Heimer, "The Unstable Alliance of Law and Morality," in Steven Hitlin and Stephen Vaisey, eds., *Handbook of the Sociology of Morality* (New York, NY: Springer, 2010), 198.

102    **"Consent" is this entire process:** Advisory Committee on Human Radiation Experiments, *Final Report of the Advisory Committee on Human Radiation Experiments* (New York: Oxford University Press, 1996), 427.

103    **Study of Infarct Survival (ISIS-2):** Major ISIS-2 results were published as ISIS-2 (Second International Study of Infarct Survival) Collaborative Group, "Randomised Trial of Intravenous Streptokinase, Oral Aspirin, both, or Neither among 17,187 Cases of Suspected Acute Myocardial Infarction: ISIS-2," *Lancet* 332, (1988): 349-360.

103    **responsibilities of the agency:** Rory Collins, Richard Doll, and Richard Peto, "Ethics of Clinical Trials," in C. J. Williams, ed., *Introducing New Treatments for Cancer: Practical, Ethical, and Legal Problems* (Chichester; New York: Wiley, 1992), 49-65.

103    **CPAP or PEEP:** Wally Carlo and Namasivayan Ambalavanan, Consent Form for the Surfactant Positive Airway Pressure and Pulse Oximetry Trial in Extremely Low Birth Weight Infants (2005), Accessed on October 13, 2022 from NICHHD at https://www.nichd.nih.gov/sites/default/files/about/Documents/2 013-04-08_All_Consent_SUPPORT.pdf.

104    **cognitively overwhelming:** Omri Ben-Shahar and Carl E. Schneider, *More than You Wanted to Know: The Failure of Mandated Disclosure* (Princeton: Princeton University Press, 2014). Ben-Shahar and Schneider achieve what I would have thought impossible—a thoroughly entertaining analysis of a topic that should be dry as dust.

104    **which were finalized in 2018:** Department of Homeland Security and 15 other departments and agencies, *Final Rule, 82 Fed Reg 7149-7273* (January 19, 2017) and Department of Homeland Security and 15 other departments and agencies, *Federal Policy for the Protection of Human Subjects: Delay of the Revisions to the Federal Policy for the Protection of Human Subjects, 83 Fed Reg 2885-94* (January 22, 2018).

105 **made this mistake:** Lynn Chaikin Epstein and Louis Lasagna, "Obtaining Informed Consent: Form Or Substance," *Archives of Internal Medicine* 123, no. 6 (1969): 682-688.

105 **brief, straightforward statement:** Louis Lasagna, "Special Subjects in Human Experimentation," *Daedalus* 98, no. 2 (1969): 449-462.

105 **small, almost negligible:** James Flory and Ezekiel Emanuel, "Interventions to Improve Research Participants' Understanding in Informed Consent for Research: A Systematic Review," *JAMA: The Journal of the American Medical Association* 292, no. 13 (2004): 1593-1601.

106 **will tell you they are not:** Paul Appelbaum, Loren Roth, and Charles Lidz deserve credit for identifying a related problem: the tendency of patients enrolled in research to believe "that the research, like the therapy they have received previously, is designed and will be executed in a manner of direct benefit to them." Paul S. Appelbaum, Loren H. Roth, and Charles Lidz, "The Therapeutic Misconception: Informed Consent in Psychiatric Research," *International Journal of Law and Psychiatry* 5, no. 3 (1982): 319-329.

106 **even more poorly:** M. H. Schaeffer, D. S. Krantz, A. Wichman, H. Masur, E. Reed, and J. K. Vinicky, "The Impact of Disease Severity on the Informed Consent Process in Clinical Research," *American Journal of Medicine* 100, no. 3 (01/01, 1996): 261-268; D. T. Penman, J. C. Holland, G. F. Bahna, G. Morrow, A. H. Schmale, L. R. Derogatis, C. L. Carnrike Jr., and R. Cherry, "Informed Consent for Investigational Chemotherapy: Patients' and Physicians' Perceptions," *Journal of Clinical Oncology* 2, no. 7 (1984): 849-855; Barrie R. Cassileth, Robert V. Zupkis, Katherine Sutton-Smith, and Vicki March. "Informed Consent—Why are its Goals Imperfectly Realized?" *The New England Journal of Medicine* 302, no. 16 (1980): 896.

106 **from liability for negligence:** 45 CFR 46.116.

106 **researcher as their client:** George J. Annas, "Reforming Informed Consent to Genetic Research," *JAMA: The Journal of the American Medical Association* 286, no. 18 (2001): 2326-2328.

106 **protect the scientist and the institution:** Christopher K. Daugherty, Donald M. Banik, Linda Janish, and Mark J. Ratain,

"Quantitative Analysis of Ethical Issues in Phase I Trials: A Survey Interview Study of 144 Advanced Cancer Patients," *IRB: Ethics & Human Research* 22, no. 3 (2000): 6-14.

## 6. Lost Lives

**108**    **Nathan's shortened form of consent:** See chapter two.

**108**    **rather than discouraging it:** Mark S. Frankel, "The Public Health Service Guidelines Governing Research Involving Human Subjects: An Analysis of the Policy-Making Process," PhD dissertation, George Washington University, 1972, 31. Other countries recognized that research oversight had the dual responsibilities of protecting subjects and encouraging sound research. As Noortje Jacobs writes, the architects of the Dutch research oversight system felt that governments had "a moral responsibility to increase both the quality and quantity of human research studies." Noortje Jacobs, *Ethics by Committee: A History of Reasoning Together about Medicine, Science, Society, and the State* (Chicago, IL: University of Chicago Press, 2022), 17.

**109**    **even if you never see them:** Julian Savulescu, "Bioethics: Why Philosophy is Essential for Progress," *Journal of Medical Ethics* 41, no. 1 (2015): 28-33.

**109**    **are usually called ethics committees:** This can cause some confusion, since in the United States the term "ethics committee" usually refers to bodies that are asked to consult on individual patient care. A typical example is the unconscious hospital patient, when the doctors feel there is no hope but the family wants to continue aggressive care.

**109**    **Australian oncologist D. R. H. Christie:** D. R. H. Christie, G. S. Gabriel, and K. Dear, "Adverse Effects of a Multicentre System for Ethics Approval on the Progress of a Prospective Multicentre Trial of Cancer Treatment: How Many Patients Die Waiting?" *Internal Medicine Journal* 37, no. 10 (Oct, 2007): 680-686.

**109**    **sixty unnecessary deaths:** David Stewart, Razelle Kurzrock, and I have presented a more complex model to estimate the cost in lives of regulatory delay of all kinds, including IRB review, FDA oversight, the Health Insurance Portability and Accountability Act, Internal

Revenue Service rules, and more. We used the example of lung cancer, and our analysis includes the benefit to both a patient who is cured of cancer and to another patient who is not cured, but who lives longer. The calculations assume that a new treatment for lung cancer increases the overall cure rate by 1%, and that patients who are not cured live three months longer. We also assume that the cumulative effect of the regulations mentioned above delays this new treatment by five years (this number is derived from the increased time to get a new drug approved compared to the 1960s). The delay caused by regulations would then cost 282,000 life-years in the US and 1.9 million worldwide. David J. Stewart, Simon N. Whitney, and Razelle Kurzrock. "Equipoise Lost: Ethics, Costs, and the Regulation of Cancer Clinical Research." *Journal of Clinical Oncology* 28, no. 17 (2010): 2925-2935.

110   **judge and jury in their own cause:** Robert Dingwall, "Turn Off the Oxygen..." *Law & Society Review* 41, no. 4 (2007): 787-795.

110   **lives that will not be saved:** Robert Dingwall, "The Ethical Case Against Ethical Regulation in Humanities and Social Science Research," *Twenty-First Century Society* 3, no. 1 (2008): 1-12.

111   **International Study of Infarct Survival:** "ISIS" today might be seen as an unfortunate nod to the Islamic State, but its actual reference is to a portion of the river Thames. As British scientist John Hampton explained, "Acronyms were a new thing, and there were in-jokes such as the Trial of Early Nifedipine Treatment in myocardial infarction that was called TRENT (the river that runs through Nottingham) and which was in progress at the same time as the second International Study of Infarct Survival, ISIS-2 (ISIS being the name given to the Thames as it passes through Oxford)." The humorous name is then reverse engineered to produce something more dignified like "Second International Study of Infarct Survival." John Hampton, "Commentary: The Need for Clinical Freedom," *International Journal of Epidemiology* 40, no. 4 (2011): 849-852.

113   **a patient who objected:** Rory Collins, Richard Doll, and Richard Peto, "Ethics of Clinical Trials," in C. J. Williams, ed., *Introducing New Treatments for Cancer: Practical, Ethical, and Legal Problems* (Chichester; New York: Wiley, 1992), 49-65.

113   **might find it "distasteful":** Collins, "Ethics of Clinical Trials," 61.

114     **doing studies in the USA:** Collins, "Ethics of Clinical Trials," 53.

114     **be considered unethical:** Noortje Jacobs, *Ethics by Committee: A History of Reasoning Together about Medicine, Science, Society, and the State* (Chicago, IL: University of Chicago Press, 2022).

114     **by a stunning 38 percent:** ISIS-2, "Randomised Trial of Intravenous Streptokinase."

114     **directly and predictably causing:** Collins, "Ethics of Clinical Trials."

115     **no less than 3,600:** Simon N. Whitney and Carl E. Schneider, "Viewpoint: A Method to Estimate the Cost in Lives of Ethics Board Review of Biomedical Research," *Journal of Internal Medicine* 269, no. 4 (Apr, 2011): 396-402. For a discussion of the cost in lives of regulatory delays overall (not just IRB review), which we estimate at tens to hundreds of thousands of life-years lost annually, see Stewart, "Equipoise Lost."

115     **Ethicist David Hunter:** David Hunter, "Is Research Ethics Regulation Really Killing People?" *Medical Journal of Australia* 202, no. 6 (2015): 338-339.

116     **Anders Ågård and colleagues:** A. Ågård, G. Hermeren, and Johan Herlitz, "Patients' Experiences of Intervention Trials on the Treatment of Myocardial Infarction: Is it Time to Adjust the Informed Consent Procedure to the Patient's Capacity?" *Heart* 86, no. 6 (2001): 632-637.

117     **"brief and concise":** Anne Gammelgaard, Peter Rossel, and Ole Steen Mortensen, "Patients' Perceptions of Informed Consent in Acute Myocardial Infarction Research: A Danish Study," *Social Science & Medicine* 58, no. 11 (2004): 2313-2324.

117     **later when you feel better:** Ågård, Hermeren, and Herlitz, "Patients' Experiences of Intervention Trials."

117     **ethicist Nancy Kass:** Nancy Kass, Ruth Faden, Rachel E. Fabi, Stephanie Morain, Kristina Hallez, Danielle Whicher, Sean Tunis, Rachael Moloney, Donna Messner, and James Pitcavage, "Alternative Consent Models for Comparative Effectiveness Studies: Views of Patients from Two Institutions," *AJOB Empirical Bioethics* 7, no. 2 (2016): 92-105.

118     **prior to trial enrollment:** Neal W. Dickert, K. A. Hendershot, C. D. Speight, and A. E. Fehr, "Patients' Views of Consent in Clinical

Trials for Acute Myocardial Infarction: Impact of Trial Design," *Journal of Medical Ethics* 43, no. 8 (Aug, 2017): 524-529.

118     **informed refusal:** Neal W. Dickert, Alex Llanos, and Habib Samady, "Re-Visiting Consent for Clinical Research on Acute Myocardial Infarction and Other Emergent Conditions," *Progress in Cardiovascular Diseases* 55, no. 3 (2012): 251-257.

118     **smaller than claimed:** Hunter, "Is Research Ethics Regulation."

## 7. The Expulsion of Ethics

120     **notebook that weighed five pounds:** On author's bookshelf.

121     **IRBs "do ethics":** Laura Jeanine Morris Stark, *Behind Closed Doors: IRBs and the Making of Ethical Research* (Chicago; London: The University of Chicago Press, 2012), 53.

121     **any system of regulation:** The term "regulation" can have many different meanings. I use it here to refer to a set of rules made and maintained by a private or public authority, such as a hospital or the federal government. IRBs constitute a hybrid system, in which rules made by the government are applied by a private body (i.e. a university or medical school committee). See Sydney Halpern, "Hybrid Design, Systemic Rigidity: Institutional Dynamics in Human Research Oversight," *Regulation & Governance* 2, no. 1 (Mar, 2008): 85-102.

122     **ethical boundaries or conditions:** Mark S. Frankel, "The Public Health Service Guidelines Governing Research Involving Human Subjects: An Analysis of the Policy-Making Process," PhD dissertation, George Washington University, 1972, 24.

122     **silent about ethics:** Office of the Surgeon General, US Public Health Service, "Surgeon General's Directives on Human Experimentation," *American Psychologist* 22, (1967): 350-355.

122     **sound professional judgment:** U.S. Department of Health, Education, and Welfare, *The Institutional Guide to DHEW Policy on Protection of Human Subjects (the Yellow Book)*, DHEW Publication No. (NIH) 72-102 (Washington, DC: Government Printing Office, 1971), iii.

123    **Black life that he knew:** Fred Gray, *The Tuskegee Syphilis Study: The Real Story and Beyond* (Montgomery, AL: NewSouth Books, 1998), 23.

123    **in the political process:** Gray, *Tuskegee Syphilis Study*, 105.

123    **weak review processes:** Harriet A. Washington, *Medical Apartheid: The Dark History of Medical Experimentation on Black Americans from Colonial Times to the Present* (New York: Doubleday, 2006); Department of Health, Education, and Welfare, *Final Report of the Tuskegee Syphilis Study Ad Hoc Advisory Panel* (Washington, DC: Government Printing Office, 1973).

124    **removed much fluff:** Writing almost thirty years later, Albert Jonsen, who was one of the commissioners, and Tom Beauchamp, who was a staff member, remember the *Belmont Report*'s authorship quite differently. See Albert R. Jonsen, "On the Origins and Future of the *Belmont Report*," in James F. Childress, Eric M. Meslin, and Harold T. Shapiro, eds, *Belmont Revisited: Ethical Principles for Research with Human Subjects* (Washington, DC: Georgetown University Press, 2005), 3-11 and Tom L. Beauchamp, "The Origins and Evolution of the *Belmont Report*," in Childress, Meslin, and Shapiro, *Belmont Revisited*, 12-25.

124    **the *Belmont Report*:** National Commission for the Protection of Human Subjects of Biomedical and Behavioral Research, *The Belmont Report: Ethical Principles and Guidelines for the Protection of Human Subjects of Research*, DHEW Publication no. (OS) 78-0012 (Washington, DC: Government Printing Office, 1978).

124    **no one of these principles:** The *Report* also discusses other topics, including the protection of potential subjects with impaired autonomy and what it saw as the important distinction between research and routine practice.

124    **the most important principle:** Bernard Schwetz, "Interview with Tom Lamar Beauchamp, PhD," in *Oral History of the Belmont Report and the National Commission for the Protection of Human Subjects of Biomedical and Behavioral Research*, 2004, accessed on October 9, 2022, https://www.hhs.gov/ohrp/education-and-outreach/-luminaries-lecture-series/belmont-report-25th-anniversary-interview-tbeacham/index.html.

125 **"a landmark":** Franklin G. Miller and Jonathan Kimmelman, "Introduction to the Special Issue on the Belmont Report," *Perspectives in Biology and Medicine* 63, no. 2 (2020): 219.

125 **a "must read":** Lydia M. Furman, "Rebuttal to the Python's Embrace," *Pediatrics* 129:3, March 2012. This comment is part of Dr. Furman's rebuttal to my essay describing the harms of IRB review to the SUPPORT infant oxygen study, discussed in chapter nine.

125 **Not every ethicist agrees:** Alexander Capron, for instance, believes that the Report "played virtually no part in the deliberations or conclusions of the President's Commission," which was the successor panel to the National Commission. Alexander M. Capron, "The Dog in the Night-Time: Or, the Curious Relationship of the Belmont Report and the President's Commission," in *Belmont Revisited: Ethical Principles for Research with Human Subjects*, ed. by James F. Childress, Eric M. Meslin, and Harold T. Shapiro (Washington, DC: Georgetown University Press, 2005), 29-40.

125 **Feminist:** Karen Lebacqz, "We Sure are Older but are We Wiser?" in Childress, Meslin, and Shapiro, *Belmont Revisited*, 99-110; Susan Sherwin, "Belmont Revisited through a Feminist Lens," in Childress, Meslin, and Shapiro, *Belmont Revisited*, 148-164.

125 **communitarian:** Barbara Prainsack and Alena Buyx, *Solidarity: Reflections on an Emerging Concept in Bioethics* (London: Nuffield Council on Bioethics, 2011).

125 **Hans Jonas's 1969 essay:** Hans Jonas, "Philosophical Reflections on Experimenting with Human Subjects," *Daedalus* 98, no. 2 (1969): 219-247.

127 **OHRP links to the *Belmont Report*:** US Department of Health and Human Services, "OHRP," expiration date August 31, 2023, accessed November 26, 2022, https://www.hhs.gov/ohrp/-index.html.

127 **mentions the *Report* not once:** Jerry Menikoff, "Where's the Law? Uncovering the Truth about IRBs and Censorship," *Northwestern University Law Review* 101, no. 2 (Feb 1, 2007): 791-799.

127 **his 1966 exposé:** Henry K. Beecher, "Ethics and Clinical Research," *The New England Journal of Medicine* 274, no. 24 (June 16, 1966): 1354-1360.

128   **James Anderson and his colleagues:** James R. Anderson, Toby L. Schonfeld, Timothy K. Kelso, and Ernest D. Prentice, "Women in Early Phase Trials: An IRB's Deliberations," *IRB: Ethics & Human Research* 25, no. 4 (2003): 7-11.

128   **first as her dissertation:** Laura Jeanine Morris Stark, "Morality in Science: How Research is Evaluated in the Age of Human Subjects Regulation," PhD dissertation, Princeton University, 2006.

128   **in the book:** Laura Jeanine Morris Stark, *Behind Closed Doors: IRBs and the Making of Ethical Research* (Chicago; London: The University of Chicago Press, 2012).

129   **cannot actually be used:** Stark, "Morality in Science," 169. Maureen Fitzgerald and colleagues, who conducted a major study of IRB deliberations in five English-speaking countries, do not even mention the *Belmont Report*. See Maureen H. Fitzgerald and Paul A. Phillips, "Centralized and Non-Centralized Ethics Review: A Five Nation Study," *Accountability in Research* 13, no. 1 (2006): 47-74; Maureen Fitzgerald, "Punctuated Equilibrium, Moral Panics and the Ethics Review Process," *Journal of Academic Ethics* 2, no. 4 (2005): 315-338; and Maureen H. Fitzgerald, Paul A. Phillips, and Elisa Yule, "The Research Ethics Review Process and Ethics Review Narratives," *Ethics & Behavior* 16, no. 4 (2006): 377-395.

129   **the enormous risk:** Caroline H. Bledsoe, Bruce Sherin, Adam G. Galinsky, Nathalia M. Headley, Carol A. Heimer, Erik Kjeldgaard, James Lindgren, Jon D. Miller, Michael E. Roloff, and David H. Uttal, "Regulating Creativity: Research and Survival in the IRB Iron Cage," *Northwestern University Law Review* 101, no. 2 (Feb 1, 2007): 593-641.

129   **requesting OHRP assistance:** Robert L. Klitzman, "Local IRBs Vs. Federal Agencies: Shifting Dynamics, Systems, and Relationships," *Journal of Empirical Research on Human Research Ethics: JERHRE* 7, no. 3 (Jul, 2012): 50-62.

130   **reactive hyperprotectionism:** Greg Koski, "Beyond Compliance...is it Too Much to Ask?" *IRB: Ethics & Human Research* 25, no. 5 (2003): 5-6.

130   **commonsense part of their jobs:** Sarah L. Babb, *Regulating Human Research: IRBs from Peer Review to Compliance Bureaucracy* (Stanford, CA: Stanford University Press, 2020), 60.

131 **individual IRB members:** See the discussion of the SUPPORT trial in chapter nine. See also Michelle M. Mello, David M. Studdert, and Troyen A. Brennan, "The Rise of Litigation in Human Subjects Research," *Annals of Internal Medicine* 139, no. 1 (2003): 40-45.

131 **struggle to attract and retain:** Norman Fost and Robert J. Levine, "The Dysregulation of Human Subjects Research," *JAMA: The Journal of the American Medical Association* 298, no. 18 (Nov 14, 2007): 2196-2198 and Institute of Medicine, *Beyond the HIPAA Privacy Rule: Enhancing Privacy, Improving Health Through Research* (Washington, DC: The National Academies Press, 2009), 12-13.

131 **actions were later challenged:** Psychiatrist Robert Klitzman has explored in depth the difficult situation of IRBs, caught between disgruntled scientists and difficult-to-please federal officials. See, for instance, Robert L. Klitzman, *The Ethics Police?: The Struggle to make Human Research Safe* (Oxford; New York: Oxford University Press, 2015) and particularly Robert L. Klitzman, "Local IRBs Vs. Federal Agencies: Shifting Dynamics, Systems, and Relationships," *Journal of Empirical Research on Human Research Ethics: JERHRE* 7, no. 3 (Jul, 2012): 50-62.

132 **makes research less ethical:** Carol A. Heimer, "The Unstable Alliance of Law and Morality," in Steven Hitlin and Stephen Vaisey, eds., *Handbook of the Sociology of Morality* (New York, NY: Springer, 2010), 197. Psychiatrist and educator Philip Candilis and colleagues provide this cogent summary: "Focused heavily on new levels of accreditation and monitoring, a new ethic of scrutiny is overtaking human experimentation." Philip J. Candilis, Rasim Arikan, Sheila B. Noone, and Jacob C. Holzer, "The New Research Ethic: Will Oversight Requirements Sink Forensic Research?" *The Journal of the American Academy of Psychiatry and the Law* 33, no. 3 (2005): 361. The same process has unfolded in other countries. Dutch scholar Noortje Jacobs reports that in 2001 "Heleen Dupuis, famous doyenne of the Dutch health ethics movement, cried out that Dutch ethics committees had turned into 'bureaucratic straightjackets' in which moral pathos was replaced by 'managerial arrogance and a mania for organization.'" Noortje Jacobs, *Ethics by Committee: A History of Reasoning Together about Medicine,*

*Science, Society, and the State* (Chicago, IL: University of Chicago Press, 2022), 172.

## 8. Evidence

135    **inevitable state of affairs:** Normal and inevitable, but not desirable when it can be avoided. See Rebecca Dresser and Joel Frader, "Off-Label Prescribing: A Call for Heightened Professional and Government Oversight," *Journal of Law, Medicine & Ethics* 37, no. 3 (2009): 476-486.

136    **it was time for a trial:** This is the concept of equipoise—the idea that two competing treatments are roughly in balance, with at least a significant minority of doctors using each, and so a study is ethical. I should add that while this view is conventional (and seems to me correct), equipoise itself has its critics. See, for instance, Franklin G. Miller and Howard Brody. "Clinical Equipoise and the Incoherence of Research Ethics." *The Journal of Medicine and Philosophy* 32, no. 2 (2007): 151-165.

136    **flecainide, encainide, or a placebo:** CAST also assigned some subjects to receive moricizine. This subgroup came to be called the CAST II trial, and it showed that moricizine was also harmful.

137    **total deaths were 26:** Debra S. Echt, Philip R. Liebson, L. Brent Mitchell, Robert W. Peters, Dulce Obias-Manno, Allan H. Barker, Daniel Arensberg, Andrea Baker, Lawrence Friedman, and H. Leon Greene, "Mortality and Morbidity in Patients Receiving Encainide, Flecainide, Or Placebo: The Cardiac Arrhythmia Suppression Trial," *New England Journal of Medicine* 324, no. 12 (1991): 781-788.

137    **the number who were killed:** Craig M. Pratt and Lemuel A. Moyé, "The Cardiac Arrhythmia Suppression Trial: Casting Suppression in a Different Light," *Circulation* 91, no. 1 (1995): 245-247.

137    **3,800 people every year:** CDC data show 760,000 heart attacks in the US in 1987. The 10% figure given antiarrhythmics is a guess; I could find no published data on this point. The 5% fatality is drawn from the CAST data.

138    **as well as a real operation:** J. Bruce Moseley, Kimberly O'Malley, Nancy J. Petersen, Terri J. Menke, Baruch A. Brody, David H. Kuykendall, John C. Hollingsworth, Carol M. Ashton, and Nelda P.

Wray, "A Controlled Trial of Arthroscopic Surgery for Osteoarthritis of the Knee," *New England Journal of Medicine* 347, no. 2 (2002): 81-88.

139 **250,000 Americans a year:** Chanu Rhee, Travis M. Jones, Yasir Hamad, Anupam Pande, Jack Varon, Cara O'Brien, Deverick J. Anderson, David K. Warren, Raymund B. Dantes, and Lauren Epstein, "Prevalence, Underlying Causes, and Preventability of Sepsis-Associated Mortality in US Acute Care Hospitals," *JAMA Network Open* 2, no. 2 (2019): e187571.

139 **PETAL:** For details, see "PETAL NETWORK, Prevention and Early Treatment of Acute Lung Injury," last updated 2022, accessed November 9, 2022, https://petalnet.org/.

139 *New York Times* **reporter:** Pam Belluck, "32 Days on a Ventilator: One Covid Patient's Fight to Breathe Again," *New York Times* 4/26/20, updated 5/8/20.

140 **half of all patients with ARDS die:** Lorraine B. Ware and Michael A. Matthay. "The Acute Respiratory Distress Syndrome." *New England Journal of Medicine* 342, no. 18 (2000): 1334-1349.

141 **far more fatalities:** ARDS Network, "Ventilation with Lower Tidal Volume as Compared with Traditional Volumes for Acute Lung Injury and the Acute Respiratory Distress Syndrome," *New England Journal of Medicine* 342, (2000): 1301-1308; Martin J. Tobin, "Culmination of an Era in Research on the Acute Respiratory Distress Syndrome," *New England Journal of Medicine* 342, no. 18 (2000): 1360-1361.

141 **in the treatment of ARDS:** Richard H. Kallet, "What is the Legacy of the National Institutes of Health Acute Respiratory Distress Syndrome Network?" *Respiratory Care* 54, no. 7 (2009): 912-924.

141 **thousands of lives every year:** This section draws heavily on the meticulous analysis in Robert Steinbrook, "How Best to Ventilate? Trial Design and Patient Safety in Studies of the Acute Respiratory Distress Syndrome," *New England Journal of Medicine* 348, no. 14 (2003): 1393-1401.

141 **how much fluid:** This study also examined the use of invasive monitoring for patients on ventilators.

141 **We have been flying blind:** Steinbrook, "How Best to Ventilate?" 1396.

143    **world class investigations:** Steinbrook, "How Best to Ventilate?" 1398.

143    **factors unique to each subject:** Steinbrook, "How Best to Ventilate?" 1399.

143    **to achieve a specific goal:** Jerry Menikoff and Edward P. Richards, *What the Doctor Didn't Say: The Hidden Truth about Medical Research* (Oxford: Oxford University Press, 2006), 33.

145    **almost entirely meaningless:** Steinbrook, "How Best to Ventilate?" 1399.

146    **not releasing their names:** Steinbrook, "How Best to Ventilate?"; OHRP letter to ARDSNet, signed by Kristina Borror and Michael Carome, addressed to Ronald Newbower and other investigators, July 3, 2003, in author's files.

## 9. Babies

148    **into the thousands:** This discussion of the development of intensive care for premature infants, of the recognition of retinopathy of prematurity, and of Silverman's experiences at Babies Hospital, along with some other material, is drawn from William A. Silverman, *Retrolental Fibroplasia: A Modern Parable* (Grune & Stratton New York, 1980).

149    **be tested in children:** Karel Allegaert, "Pediatric Clinical Pharmacology: An Introduction to a Series of Educational Papers," *European Journal of Pediatrics* 172: (2013) 289-292.

150    **The best of these:** Fifty years later, Ann Stark was lead author on a Neonatal Research Network study that showed that using steroids in the hope of protecting premature babies from lung disease produced no pulmonary benefit and led to decreased growth and an increased chance of intestinal perforation. Ann R. Stark, Waldemar A. Carlo, Jon E. Tyson, Lu-Ann Papile, Linda L. Wright, Seetha Shankaran, Edward F. Donovan, William Oh, Charles R. Bauer, and Shampa Saha, "Adverse Effects of Early Dexamethasone Treatment in Extremely-Low-Birth-Weight Infants," *New England Journal of Medicine* 344, no. 2 (2001): 95-101.

151    **Arnall Patz:** Silverman, *Retrolental Fibroplasia*.

**152**    **Oxygen was blinding babies:** A. Patz, L.E. Hoeck, and E. De La Cruz, "Studies on the Effect of High Oxygen Administration in Retrolental Fibroplasia. I. Nursery Observations," *American Journal of Ophthalmology* 35, no. 9 (Sep, 1952): 1248-1253.

**152**    **oxygen was leaking out:** Jon Tyson described the Tulane experience, and its leaky incubators, in a conversation with me.

**153**    **Patrick Bouvier Kennedy:** Lawrence K. Altman, "A Kennedy Baby's Life and Death," *New York Times,* July 29, 2013.

**153**    **would be home in a week:** Conversation with Jon Tyson.

**153**    **the remaining third:** A. Dance, "Survival of the Littlest: The Long-Term Impacts of being Born Extremely Early," *Nature* 582, no. 7810 (Jun, 2020): 20-23.

**153**    **improving over time:** Jon E. Tyson, Nehal A. Parikh, John Langer, Charles Green, and Rosemary D. Higgins, "Intensive Care for Extreme Prematurity—Moving Beyond Gestational Age," *New England Journal of Medicine* 358, no. 16 (2008): 1672-1681.

**155**    **The regulations recognize:** The regulation in effect at the time was 45 CFR 46.116(d); the current version is 45 CFR 46.116(f).

**156**    **pure oxygen is toxic:** V. S. Kapadia, L. F. Chalak, J. E. Sparks, J. R. Allen, R. C. Savani, and M. H. Wyckoff, "Resuscitation of Preterm Neonates with Limited Versus High Oxygen Strategy," *Pediatrics* 132, no. 6 (Dec, 2013): e1488-96.

**156**    **to this day, unknown:** William A. Silverman, "A Cautionary Tale about Supplemental Oxygen: The Albatross of Neonatal Medicine," *Pediatrics* 113, no. 2 (Feb, 2004): 394-396.

**157**    **the published literature:** C. H. Cole, K. W. Wright, W. Tarnow-Mordi, D.L. Phelps, and the Pulse Oximetry Saturation Trial for Prevention of Retinopathy of Prematurity Planning Study Group. "Resolving our Uncertainty about Oxygen Therapy." *Pediatrics* 112, no. 6 Pt 1 (Dec, 2003): 1415-1419.

**158**    **to try both levels:** Cole, "Resolving our Uncertainty."

**159**    **also better survival:** Waldemar A. Carlo, Edward F. Bell, and Michele C. Walsh, "Oxygen-Saturation Targets in Extremely Preterm Infants," *New England Journal of Medicine* 368, no. 20 (2013): 1949-1950. Blindness was 8.6% in the low-oxygen group and 17.9% in the high-oxygen group.

159    **infants born during that delay:** Wade D. Rich, N. N. Finer, M. G. Gantz, N. S. Newman, A. M. Hensman, E. C. Hale, K. J. Auten, et al, "Enrollment of Extremely Low Birth Weight Infants in a Clinical Research Study may Not be Representative," *Pediatrics* 129, no. 3 (Mar, 2012): 480-484; Wade D. Rich, Kathy J. Auten, Marie G. Gantz, Ellen C. Hale, Angelita M. Hensman, Nancy S. Newman, and Neil N. Finer, "Antenatal Consent in the SUPPORT Trial: Challenges, Costs, and Representative Enrollment," *Pediatrics (Evanston)* 126, no. 1 (Jul, 2010): e215-e221; Simon N. Whitney, "The Python's Embrace: Clinical Research Regulation by Institutional Review Boards," *Pediatrics* 129, no. 3 (2012): 576-578.

159    **harmed by being** *excluded***:** Rebecca Dresser, *When Science Offers Salvation: Patient Advocacy and Research Ethics* (Oxford; New York: Oxford University Press, 2001), 55.

159    **women and minorities:** Giselle Corbie-Smith, "The Continuing Legacy of the Tuskegee Syphilis Study: Considerations for Clinical Investigation," *The American Journal of the Medical Sciences* 317, no. 1 (1999): 5-8; Sam S. Oh, Joshua Galanter, Neeta Thakur, Maria Pino-Yanes, Nicolas E. Barcelo, Marquitta J. White, Danielle M. de Bruin, Ruth M. Greenblatt, Kirsten Bibbins-Domingo, and Alan HB Wu, "Diversity in Clinical and Biomedical Research: A Promise Yet to be Fulfilled," *PLoS Medicine* 12, no. 12 (2015): e1001918; Susan Sherwin, "Belmont Revisited through a Feminist Lens," in *Belmont Revisited: Ethical Principles for Research with Human Subjects*, eds. James F. Childress, Eric M. Meslin, and Harold T. Shapiro (Washington, DC: Georgetown University Press, 2005), 148-164.

159    **instructed investigators:** National Institutes of Health, "NIH Inclusion Outreach Toolkit: How to Engage, Recruit, and Retain Women in Clinical Research," n.d., accessed April 25, 2020, https://orwh.od.nih.gov/toolkit/recruitment/history.

159    **inclusive enrollment mandatory:** National Institutes of Health Revitalization Act of 1993, Public Law 103-43, 103d Congress, June 10, 1993.

160    **too little time:** Rich *et al*, "Antenatal Consent in the SUPPORT Trial."

160 **children of relative privilege:** Rich et al., "Enrollment of Extremely Low Birth Weight Infants"; Rich et al., "Antenatal Consent in the SUPPORT Trial"; Whitney, "The Python's Embrace."

160 **letter of reprimand:** Lisa R. Buchanan, Division of Compliance Oversight, OHRP, letter to Richard B. Marchase, V.P. for Research and Development, University of Alabama at Birmingham, March 7, 2013.

161 **even if not participating:** 45 CFR 46.111(a)(2).

162 **generally did at least as well:** John D. Lantos, "The Weird Divergence of Ethics and Regulation with Regard to Informed Consent," *The American Journal of Bioethics* 13, no. 12 (2013): 31-33.

163 **the *Times* editorial board:** "An Ethical Breakdown," editorial, *New York Times*, April 15, 2013.

163 **Dallas researchers fail to explain:** Sherry Jacobson, "Dallas Researchers Fail to Explain Why Parents in National Baby Study Weren't Notified of Fatal Risks," *Dallas Morning News,* April 11, 2013.

163 **17 of them died:** Sherry Jacobson, "UTSW Enrolled 73 Premature Infants in Controversial Study; 17 of them Died," *Dallas Morning News,* April 13, 2013.

164 **would have studied something else:** Jon Tyson, Michele Walsh, and Carl D'Angio have written a lucid summary of the mistaken assumptions behind many of the criticisms of the SUPPORT trial. See Jon E. Tyson, Michele Walsh, and Carl T. D'Angio, "Comparative Effectiveness Trials: Generic Misassumptions Underlying the SUPPORT Controversy," *Pediatrics* 134, no. 4 (2014): 651-654.

164 **Public Citizen:** Michael A. Carome, Public Citizen letter to Secretary Sebelius about the SUPPORT Trial, April 10, 2013.

164 **Alliance for Human Research Protection:** Vera Sharav, "An Ethical Breakdown," Alliance for Human Research Protection, April 16, 2013, https://ahrp.org/an-ethical-breakdown, accessed February 13, 2021.

164 **attracted a team of lawyers:** *Looney et al v. Moore et al,* Case Number 2:2013cv00733, Southern Office, US District Court for the Northern District of Alabama, 2013.

164 **put other research on hold:** Kathy L. Hudson, Alan E. Guttmacher, and Francis S. Collins. "In Support of SUPPORT—a View

from the NIH." *New England Journal of Medicine* 368, no. 25 (2013): 2349-2351.

165    **that sharply criticized OHRP:** Hudson, Guttmacher, and Collins, "In Support of SUPPORT."

165    **befuddled IRBs:** The controversy led to two letters to the *New England Journal of Medicine*, one supporting OHRP's action, the other attacking it, each signed by a long list of ethicists and pediatricians. See Benjamin S. Wilfond, David Magnus, Armand H. Antommaria, Paul Appelbaum, Judy Aschner, Keith J. Barrington, Tom Beauchamp, Renee D. Boss, Wylie Burke, and Arthur L. Caplan, "The OHRP and SUPPORT," *New England Journal of Medicine* (2013) (applauding the SUPPORT investigators) and Ruth Macklin, Lois Shepherd, Alice Dreger, Adrienne Asch, Francoise Baylis, Howard Brody, Larry R. Churchill, Carl H. Coleman, Ethan Cowan, and Janet Dolgin, "The OHRP and SUPPORT—Another View," *New England Journal of Medicine* 369, no. 2 (2013): e3 (praising OHRP).

165    **The agency responded:** Lisa R. Buchanan, Division of Compliance Oversight, OHRP, Letter to Richard B. Marchase, V.P. for Research & Economic Development, University of Alabama at Birmingham, June 4, 2013.

165    **meeting to discuss SUPPORT:** Department of Health and Human Services, "Public Meeting: Matters Related to Protection of Human Subjects and Research Considering Standard of Care Interventions," August 28, 2013, https://www.hhs.gov/ohrp/regulations-and-pol-icy/requests-for-comments/public-Meeting-08-28-2013/index.-Html, accessed February 14, 2021.

165    **increased risk of death:** George J. Annas and Catherine L. Annas, " 'Unusual Care': Groupthink and Willful Blindness in the SUPPORT Study," *The American Journal of Bioethics* 20, no. 1 (2020): 44-46.

166    **misleading and dangerous:** John Lantos, comments at Department of Health and Human Services, "Public Meeting." See also John D. Lantos, "OHRP and Public Citizen are Wrong about Neonatal Research on Oxygen Therapy," *Bioethics Forum* 2020, no. 11/16 (2013) and John D. Lantos, "Learning the Right Lessons from the SUPPORT Study Controversy," *Archives of Disease in Childhood, Fetal and Neonatal Edition* 99, no. 1 (Jan, 2014): F4-5.

**166**   **a subject of an experiment:** Shawn Pratt, comments at Department of Health and Human Services, "Public Meeting."

**166**   **the best treatment is unknown:** See Susan H. Wootton, Patricia W. Evans, and Jon E. Tyson. "Unproven Therapies in Clinical Research and Practice: The Necessity to Change the Regulatory Paradigm." *Pediatrics* 132, no. 4 (2013): 599-601. Tyson made the same argument almost twenty years before, arguing that every child receiving treatments that had not been carefully evaluated was participating in the experiment that is often the true nature of routine practice; see Jon E. Tyson, "Use of Unproven Therapies in Clinical Practice and Research: How can we Better Serve our Patients and their Families?" *Seminars in Perinatology* 19 no. 2 (1995) 98-111.

**166**   **babies not enrolled:** Jon Tyson, comments at Department of Health and Human Services, "Public Meeting."

**167**   **complication short of death:** Rich et al., "Enrollment of Extremely Low Birth Weight Infants."

**167**   **can be attributed to SUPPORT?** Veda Sharav, statement at DHHS meeting.

**168**   **16.2 percent, 19.9 percent, and 24.1percent:** SUPPORT Study Group of the Eunice Kennedy Shriver NICHD Neonatal Research Network, "Early CPAP Versus Surfactant in Extremely Preterm Infants," *New England Journal of Medicine* 362, no. 21 (2010): 1970-1979; SUPPORT Study Group of the Eunice Kennedy Shriver NICHD Neonatal Research Network, "Target Ranges of Oxygen Saturation in Extremely Preterm Infants," *New England Journal of Medicine* 362, no. 21 (2010): 1959-1969.

**168**   **who were not in the study:** Lantos, "Weird Divergence."

**169**   **Vindication for SUPPORT:** John D. Lantos, "Vindication for SUPPORT," *New England Journal of Medicine* 373, no. 15 (2015): 1393-1395.

**169**   **doesn't change that:** Sabrina Tavernise, "Premature Babies Study Raises Debate Over Risks and Ethical Consent," *New York Times,* September 7, 2015.

**172**   **a personal viewpoint:** John Lantos and Jerry Menikoff discussion at DHHS meeting.

**172**   **Sharissa Cook said:** Statement at DHHS meeting.

**173**   **Shawn Pratt said:** Statement at DHHS meeting.

173    **not considered lying?** Carrie Pratt, statement at Public Citizen Press Conference about SUPPORT Study, August 28, 2031, Washington, D.C. The press conference, in three videos, was accessed on January 14, 2021, at https://www.Youtube.com/watch?-v=BcjZce0eM_8; https://www.Youtube.com/watch?v=ommX3lqm-VXw; https://www.Youtube.com/watch?v=trg5ry8FJuU&t=423s.

## 10. Failure

175    **just five subjects:** The best-known was Jesse Gelsinger, who suffered a fatal reaction during an attempt at gene transfer. This tragic loss of a lovable young man with an engaging personality led to hundreds of articles in both the professional and lay media, often featuring photographs of Gelsinger as an adolescent. The field of gene therapy was brought to a halt for a decade. Thankfully, cases like this are extremely rare.

175    **medical research very safe:** Steven Joffe, "Revolution Or Reform in Human Subjects Research Oversight," *Journal of Law, Medicine & Ethics* 40, no. 4 (2012): 922-929.

176    **spent about $1 million:** Jeanne L. Speckman, Margaret M. Byrne, Jason Gerson, Kenneth Getz, Gary Wangsmo, Carianne T. Muse, and Jeremy Sugarman, "Determining the costs of institutional review boards," *IRB: Ethics & Human Research* 29, no. 2 (2007): 7-13; Norman Fost, "Fusing and Confusing Ethics and Regulations," in Paula Knudson, Bridget Gardner and Joan Rachlin, eds., *PRIM&R Through the Years: Three Decades of Protecting Human Subjects* (Boston, MA: Public Responsibility in Medicine and Research, 2006), 457.

176    **$40 billion on medical research:** NIH, "Budget," https://www.nih.gov/about-nih/what-we-do/budget, reviewed 6/29/20, accessed 8/18/21. The NIH does not provide a specific amount for human subject research, but in 2011 the Presidential Commission estimated it at about half of the total. Presidential Commission for the Study of Bioethical Issues, *Moral Science: Protecting Participants in Human Subjects Research* (Washington, D.C., 2011), 34. These numbers exclude the value of time volunteered by IRB members.

176 **vastly more suffering and death:** Carol A. Heimer, "The Unstable Alliance of Law and Morality," in Steven Hitlin and Stephen Vaisey, eds., *Handbook of the Sociology of Morality* (New York, NY: Springer, 2010), 179-202.

177 **people aged 35 to 54:** Sameer Arora, George A. Stouffer, Anna M. Kucharska-Newton, Arman Qamar, Muthiah Vaduganathan, Ambarish Pandey, Deborah Porterfield, Ron Blankstein, Wayne D. Rosamond, and Deepak L. Bhatt, "Twenty Year Trends and Sex Differences in Young Adults Hospitalized with Acute Myocardial Infarction: The ARIC Community Surveillance Study," *Circulation* 139, no. 8 (2019): 1047-1056.

178 **no voices of dissent:** Will C. van den Hoonaard, *The Seduction of Ethics: Transforming the Social Sciences* (Toronto: University of Toronto Press, 2011). As Carl Elliott writes of bioethics more generally, experts "are acculturated into a standard view of the world, conformity to which is rewarded through promotions, pay raises, awards, and increased status in the field. See Carl Elliott, "The Tyranny of Expertise," in *The Ethics of Bioethics: Mapping the Moral Landscape*, ed. Lisa A. Eckenwiler and Felicia G. Cohn (Baltimore: Johns Hopkins University Press, 2007), 46.

179 **self-protective ideology:** Howard S. Becker, "Comment on Kevin D. Haggerty, 'Ethics Creep: Governing Social Science Research in the Name of Ethics'," *Qualitative Sociology* 27, no. 4 (2004): 415-416. The story of the IRB system's overreach into the social sciences is deeply troubling—but, for us, it's a story for another day.

179 **comparison to Nazi doctors:** Scott Burris, "Regulatory Innovation in the Governance of Human Subjects Research: A Cautionary Tale and some Modest Proposals," *Regulation & Governance* 2, no. 1 (Mar, 2008): 65-84.

179 ***in terrorem* effect:** Dale Carpenter, "Institutional Review Boards, Regulatory Incentives, and some Modest Proposals for Reform," *Northwestern University Law Review* 101, no. 2 (Feb 1, 2007): 687. William Banner has made the same point. See William Banner, " 'It is Easier to Denature Plutonium than to Denature the Evil Spirit of Man'--Albert Einstein," *Clinical Toxicology* 50, no. 7 (2012): 537.

180 **veiled in procedural details:** Fredric L. Coe, "Costs and Benefits of a Well-Intended Parasite: A Witness and Reporter on the IRB

Phenomenon," *Northwestern University Law Review* 101, (2007): 723-733.

**180** **lack of respect:** Robert J. Amdur and Elizabeth A. Bankert. *Institutional Review Board: Member Handbook*, 3rd ed. Sudbury, MA.: Jones and Bartlett, 2011, 42. There is no official IRB handbook. Amdur and Bankert's is, however, quite popular.

**180** **requirements for revision:** Levine, Robert J. *Ethics and Regulation of Clinical Research*, 2nd ed. (Baltimore: Urban & Schwarzenberg, 1986), 333.

**180** **not to accept:** The regulations require an IRB to explain to the investigator why it has disapproved a proposed plan of research, but no explanation is required when the committee simply requires changes. An IRB that wants to scuttle a proposal could therefore require, for instance, that enrollment is capped at a level at which significant results are unlikely, or that a pilot study be done first, knowing that there will be too little time to complete the actual project.

**181** **the final changes:** Department of Homeland Security and 15 other departments and agencies, *Final Rule, 82 Fed Reg 7149-7273* (January 19, 2017); Department of Homeland Security and 15 other departments and agencies, *Federal Policy for the Protection of Human Subjects: Delay of the Revisions to the Federal Policy for the Protection of Human Subjects, 83 Fed Reg 2885-94* (January 22, 2018). It will take years for OHRP and IRBs to reach a new equilibrium as they implement these complex modifications. See P. Pearl O'Rourke, "The Final Rule: When the Rubber Meets the Road," *The American Journal of Bioethics* 17, no. 7 (2017): 27-33.

**181** **regulatory burden little changed:** As a research administrator interviewed by Sarah Babb complained, "All they did was tinker around the edges." Sarah L. Babb, *Regulating Human Research: IRBs from Peer Review to Compliance Bureaucracy* (Stanford, CA: Stanford University Press, 2020), 107. Oncologists Razelle Kurzrock and David Stewart liken the impact of regulatory burden (including but not limited to IRB requirements) to the mud in World War I trench warfare, saying it is "onerous, misguided, and expensive, with little value added." David J. Stewart and Razelle Kurzrock, "Cancer:

The Road to Amiens," *Journal of Clinical Oncology* 27, no. 3 (Jan 20, 2009): 328-333.

181   **Coe tells me:** Email from Fredric Coe, March 1, 2020.

182   **American Association of University Professors:** American Association of University Professors, "Institutional Review Boards and Social Science Research," 2000, accessed October 19, 2022, https://www.aaup.org/report/institutional-review-boards-and-social-science-research. David Hyman, who served on the AAUP committee that prepared the report, presents an excellent review of the issues in David A. Hyman, "The Pathologies of Institutional Review Boards: Are IRBs the 'Least Worst' Way to Promote Research Ethics?" *Regulation* 30, (2007): 42-49 and David A. Hyman, "Institutional Review Boards: Is this the Least Worst we can do?" *Northwestern University Law Review* 101, no. 2 (Feb 1, 2007): 749.

182   **substantially modified:** IOM (Institute of Medicine), *Beyond the HIPAA Privacy Rule: Enhancing Privacy, Improving Health Through Research* (Washington, DC: The National Academies Press, 2009).

182   **deep-seated resistance:** I. Glenn Cohen, who is Director of Harvard Law School's Petrie-Flom Center for Health Law Policy, Biotechnology, and Bioethics has written that, while it is easy to propose fundamental changes in IRB review, "There is a huge medico-industrial complex that supports and benefits from the current paradigms in human subjects research that would resist many of these changes. Moreover there is a set of understandably risk-averse governmental officers who do not want to be known as the ones who let "the next Tuskegee" happen." I. Glenn Cohen, "Introduction to Part V—Paradigm Shifts in Research Ethics," in I. Glenn Cohen and Holly Fernandez Lynch, eds, *Human Subjects Research Regulation: Perspectives on the Future* (Cambridge, MA: MIT Press, 2014), 284.

182   **to prevent a repetition:** Serious scholars of the abuse of Black bodies by white doctors are at pains to point out that Tuskegee was only the most egregious in a long, long line of abuses. See Harriet A. Washington, *Medical Apartheid: The Dark History of Medical Experimentation on Black Americans from Colonial Times to the Present* (New York: Doubleday, 2006); W. Michael Byrd and Linda A. Clayton, *An American Health Dilemma, Volume I: A Medical History*

*of African Americans and the Problem of Race, Beginnings to 1900* (New York: Routledge, 2000); W. Michael Byrd and Linda A. Clayton, *An American Health Dilemma, Volume II: Race, Medicine, and Health Care in the United States 1900–2000* (New York: Routledge, 2002); Ralph V. Katz and Rueben C. Warren, eds, *The Search for the Legacy of the USPHS Syphilis Study at Tuskegee* (Lanham, MD: Lexington Books, 2011); Harold L. Aubrey, "African Americans and the Broader Legacy of Experience with the American Health Care Community: Parasites, Locusts and Scavengers," in Ralph V. Katz and Rueben C. Warren, eds, *The Search for the Legacy of the USPHS Syphilis Study at Tuskegee* (Lanham, MD: Lexington Books, 2011), 107-115.

182    **the CDC website says:** Centers for Disease Control and Prevention, "U.S. Public Health Service Syphilis Study at Tuskegee: Research Implications—How Tuskegee Changed Research Practices," updated 2020, accessed April 28, 2020, https://www.cdc.gov/tuskegee/after.htm.

183    **generalizability of their results:** Byrd and Clayton, *An American Health Dilemma, Volume II*, 572.

183    **including minority subjects:** Vivian Pinn, "From Exclusion to Inclusion—Participation in Biomedical Research and the Legacy of the Public Health Syphilis Study at Tuskegee," in Ralph V. Katz and Rueben C. Warren, eds, *The Search for the Legacy of the USPHS Syphilis Study at Tuskegee* (Lanham, MD: Lexington Books, 2011), 1-2, 16.

183    **'too much protection':** Giselle Corbie-Smith, "The Continuing Legacy of the Tuskegee Syphilis Study: Considerations for Clinical Investigation," *The American Journal of the Medical Sciences* 317, no. 1 (1999): 5-8.

183    **fording the gulf that yawns:** Washington, *Medical Apartheid*, 387.

183    **like Tuskegee in reverse:** Ruth E. Malone, Valerie B. Yerger, Carol McGruder, and Erika Froelicher, "'It's Like Tuskegee in Reverse': A Case Study of Ethical Tensions in Institutional Review Board Review of Community-Based Participatory Research," *American Journal of Public Health* 96, no. 11 (2006): 1914-1919.

183 **fear of the past:** As Amy Fairchild and Ronald Bayer have commented, "While Tuskegee can stimulate productive reflection on questions of social justice, its reckless invocation risks derailing serious and sustained discussion of the dilemmas posed by research with vulnerable populations." Amy L. Fairchild and Ronald Bayer, "Uses and Abuses of Tuskegee," *Science* 284, no. 5416 (1999): 919-921.

184 **agree is broken:** Laura Jeanine Morris Stark, "IRBs and the problem of 'local precedents'," in I. Glenn Cohen and Holly Fernandez Lynch, eds, *Human Subjects Research Regulation: Perspectives on the Future* (Cambridge, MA: MIT Press, 2014), 173-186.

184 **necessary and long overdue:** Greg Koski, "Getting Past Protectionism: Is it Time to Take Off the Training Wheels?" in I. Glenn Cohen and Holly Fernandez Lynch, eds, *Human Subjects Research Regulation: Perspectives on the Future* (Cambridge, MA: MIT Press, 2014), 346.

185 **from the ground up:** Rosamond Rhodes, Jody Azzouni, Stefan Bernard Baumrin, Keith Benkov, Martin J. Blaser, Barbara Brenner, Joseph W. Dauben et al, "*De minimis* Risk: A Proposal for a New Category of Research Risk," *The American Journal of Bioethics* 11, no. 11 (2011): 1-7.

185 **new ethics framework:** Ruth R. Faden, Nancy E. Kass, Steven N. Goodman, Peter Pronovost, Sean Tunis, and Tom L. Beauchamp, "An Ethics Framework for a Learning Health Care System: A Departure from Traditional Research Ethics and Clinical Ethics," *The Hastings Center Report* 43, no. s1 (Jan, 2013): S16-S27.

185 **five fundamental reforms:** Ezekiel J. Emanuel, Anne Wood, Carianne Tucker Muse, Jeremy Sugarman, Alan Fleischman, Angela Bowen, Kenneth A. Getz, et al, "Oversight of Human Participants Research: Identifying Problems to Evaluate Reform Proposals," *Annals of Internal Medicine* 141, no. 4 (Aug 17, 2004): 282-291.

185 **a few of the many:** Robert L. Klitzman, *The Ethics Police?: The Struggle to make Human Research Safe* (Oxford; New York: Oxford University Press, 2015) spends two chapters on proposed changes, about a dozen each for local and federal review. Overall there are certainly hundreds, and perhaps thousands, of proposals for change at some level published every year. Carl E. Schneider, *The Censor's*

*Hand: The Misregulation of Human-Subject Research* (Cambridge, MA: MIT Press, 2015), presents the simplest solution—abolish the system—but this idea has not gained traction.

185   **is basically a failure:** Greg Koski, "Changing the Paradigm: New Directions in Federal Oversight of Human Research," *Journal of Pediatric Gastroenterology and Nutrition* 37 Supplement 1 (Nov, 2003): S2-S6. In 2015, British ethicist Julian Savulescu argued that "both bioethics and medical ethics together have, in many ways, failed as fields." This includes IRB functioning. He concluded that what was needed was better philosophy so that the ethics would be better as well. See Julian Savulescu, "Bioethics: Why Philosophy is Essential for Progress," *Journal of Medical Ethics* 41, no. 1 (2015): 28-33.

185   **agree on a diagnosis:** Psychologist Jonathan Baron believes that "those who make and apply regulations in the name of bioethics usually do not have training in decision analysis or statistics, and the philosophical principles they apply are usually not subject to the kinds of critical analysis that we would expect to see in the major academic philosophy journals. The professional field of bioethics needs an academic upgrade." Jonathan Baron, "Some Fallacies of Human-Subjects Protection, and some Solutions," *Cortex; a Journal Devoted to the Study of the Nervous System and Behavior* 65, (2014): 246-254.

185   **in how to manage risk:** Peter L. Bernstein, *Against the Gods: The Remarkable Story of Risk* (New York: John Wiley & Sons, 1996); Paul Hopkin, *Fundamentals of Risk Management,* 4th ed (New York, NY: Kogan Page, 2017).

186   **attempt at self-protection:** Nancy M. P. King, "Key Information in the New Common Rule: Can it Save Research Consent?" *Journal of Law, Medicine & Ethics* 47, no. 2 (2019): 203-212.

186   **protection of research participants:** Michael McDonald, Susan Cox, and Anne Townsend, "Toward Human Research Protection that is Evidence Based and Participant Centered," in I. Glenn Cohen and Holly Fernandez Lynch, eds, *Human Subjects Research Regulation: Perspectives on the Future* (Cambridge, MA: MIT Press, 2014).

186   **also protects the institution:** Simon N. Whitney, "A Fern in Amber: Risk Management in Research with Humans," *Risk Management* 17, no. 4 (2015): 226-239; Simon N. Whitney,

"Institutional Review Boards: A Flawed System of Risk Management," *Research Ethics* 12, no. 4 (2016): 182-200.

188  **The Church of Jesus Christ:** The Church of Jesus Christ of Latter-Day Saints, "Risk Management Division Services," n.d., accessed on November 15, 2022, https://www.churchofjesuschrist.org/study/-manual/safety-health-and-environmental-manual/01/1-3-risk-management-division-services?lang=eng; http://www.resourcepart-nersonline.org/.

188  **The Quakers:** Friends Service Alliance, "Compliance and Risk Management Partners," n.d., accessed on November 15, 2022, https://www.fsainfo.org/compliance-partners.

188  **NASA shares:** NASA, "Risk Management," updated September 18, 2022, accessed November 15, 2022, https://sma.nasa.gov/sma-disciplines/risk-management.

188  **more harm to health:** Stephen Breyer, *Breaking the Vicious Circle: Toward Effective Risk Regulation* (Cambridge, MA: Harvard University Press, 1993), 23.

189  **prevented siloed thinking:** D. H. Cowan, "Human Experimentation: The Review Process in Practice," *Case Western Reserve Law Review* 25, no. 3 (1975): 533-564.

189  **lower-ranking faculty:** Cowan, "Human Experimentation."

190  **responsibilities down the line:** Committee of Sponsoring Organizations of the Treadway Commission, (COSO), *Enterprise Risk Management—Integrating with Strategy and Performance, Executive Summary,* 2017, accessed on October 18, 2022, https://www.coso.org/sitepages/guidance-on-enterprise-risk-management.aspx?web=1.

190  **risks and opportunities:** Neil A. Doherty, *Integrated Risk Management Techniques and Strategies for Managing Corporate Risk* (New York: McGraw-Hill, 2000), 270.

190  **ill-informed or poorly-reasoned:** 45 CFR 46.112.

191  **research could not be conducted:** Comments of anonymous peer reviewer; on file with author.

## 11. Reform

192     **complains that IRB review:** Fredric L. Coe, "Costs and Benefits of a Well-Intended Parasite: A Witness and Reporter on the IRB Phenomenon," *Northwestern University Law Review* 101, (2007): 723-733.

193     **who have had kidney stones:** Data from U.S. Census and Charles D. Scales, Jr., Alexandria C. Smith, Janet M. Hanley, Christopher S. Saigal, and the Urologic Diseases in America Project, "Prevalence of Kidney Stones in the United States," *European Urology* 62, no. 1 (2012): 160-165.

194     **might be worth approving:** Nir Eyal and Tobias Gerhard, "Do Coronavirus Vaccine Challenge Trials have a Distinctive Generalisability Problem?" *Journal of Medical Ethics* 48, no. 9 (2022): 586-589; Abie Rohrig and Nir Eyal, "A New Day for Human Challenge Trials?" *Trends in Molecular Medicine* 28:7 (2022): 531-532.

194     **Novavax enrolled 30,000 people:** National Institutes of Health, "U.S. Clinical Trial Results show Novavax Vaccine is Safe and Prevents COVID-19," July 14, 2021, accessed August 17, 2021, https://www.nih.gov/news-events/news-releases/us-clinical-trial-results-show-novavax-vaccine-safe-prevents-covid-19.

195     **Rutgers ethicist Nir Eyal:** Nir Eyal, Marc Lipsitch, and Peter G. Smith, "Human Challenge Studies to Accelerate Coronavirus Vaccine Licensure," *The Journal of Infectious Diseases* 221, no. 11 (2020): 1752-1756.

195     **disclosure next to impossible:** Jeffrey P. Kahn, Leslie Meltzer Henry, Anna C. Mastroianni, Wilbur H. Chen, and Ruth Macklin, "For Now, It's Unethical to use Human Challenge Studies for SARS-CoV-2 Vaccine Development," *Proceedings of the National Academy of Sciences* 117, no. 46 (2020): 28538-28542.

196     **inviolable prohibitions and taboos:** Hans Jonas, "Philosophical Reflections on Experimenting with Human Subjects," *Daedalus* 98, no. 2 (1969): 219-247.

196     **serious injury or death:** Nir Eyal has argued persuasively (and mathematically) that we can't set a general cap on risk, since unusual circumstances justify unusual measures. See Nir Eyal, "Is There an

Ethical Upper Limit on Risks to Study Participants?" *Public Health Ethics* 13:2 (2020): 143-156.

196 **purchased insurance:** Sydney A. Halpern, *Lesser Harms: The Morality of Risk in Medical Research* (Chicago: University of Chicago Press, 2004).

196 **such a system in place:** Holly Fernandez Lynch, "Protecting Human Research Subjects as Human Research Workers," in I. Glenn Cohen and Holly Fernandez Lynch, eds, *Human Subjects Research Regulation: Perspectives on the Future* (Cambridge, MA: MIT Press, 2014), 327-340.

197 **not worth the time:** He summarizes their usual reaction: "Doc, let's get by all the words ..." Coe, "Costs and Benefits." (See chapter five).

197 **I agree to the terms:** Omri Ben-Shahar and Carl E. Schneider, *More Than You Wanted to Know: The Failure of Mandated Disclosure* (Princeton NJ: Princeton University Press, 2014).

198 **the best information they could get:** Jonathan Baron, *Against Bioethics* (Cambridge, MA: MIT Press, 2006), 119.

198 **in treating diverse patients:** Giselle Corbie-Smith, "The Continuing Legacy of the Tuskegee Syphilis Study: Considerations for Clinical Investigation," *The American Journal of the Medical Sciences* 317, no. 1 (1999): 5-8; Rebecca Dresser, *When Science Offers Salvation: Patient Advocacy and Research Ethics* (Oxford; New York: Oxford University Press, 2001); Patricia A. King, "Justice Beyond *Belmont*," in *Belmont Revisited: Ethical Principles for Research with Human Subjects*, ed. James F. Childress, Eric M. Meslin and Harold T. Shapiro (Washington, DC: Georgetown University Press, 2005), 136-147.

200 **skim or skip it:** Consent forms today are often constructed or administered so that skimming or skipping is difficult or impossible. An administrator may read every word aloud to the subject, doing their best to make sure the subject's attention does not wander, or an electronic form may require that every word be displayed and that multiple boxes affirming each important fact be individually checked.

201 **balance disorders, and bone loss:** For an interesting exploration of the health risks of long duration spaceflight, see Institute of Medicine, *Health Standards for Long Duration and Exploration*

*Spaceflight: Ethics Principles, Responsibilities, and Decision Framework* (Washington, DC: The National Academies Press, 2014).

201    **predispose them to:** Scott M. Smith, Martina Heer, Linda C. Shackelford, Jean D. Sibonga, Jordan Spatz, Robert A. Pietrzyk, Edgar K. Hudson, and Sara R. Zwart, "Bone Metabolism and Renal Stone Risk during International Space Station Missions," *Bone (New York, N.Y.)* 81, (2015): 712-720.

203    **"disqualified" from research:** George Annas, "Old and Emerging Bioethical Issues in Research on Atoms and Genes," *Protecting Human Subjects, U.S. Department of Energy, Office of Biological and Environmental Research*, Winter (1996): 10.

203    **less likely to participate:** W. Terry, L. G. Olson, P. Ravenscroft, L. Wilss, and G. Boulton-Lewis, "Hospice Patients' Views on Research in Palliative Care," *Internal Medicine Journal* 36, no. 7 (Jul, 2006): 406-413. This work was conducted in Australia; other studies, in other countries, have reached similar findings. Canadian sociologist Amanda van Beinum and her coauthors, for instance, interviewed the family members of hospice patients that had participated in research that was not intended to help them during the dying process. These families felt that this research "offered families of the dying an opportunity to affirm the intrinsic value of patients' lives and contributions." Amanda van Beinum, Nick Murphy, Charles Weijer, Vanessa Gruben, Aimee Sarti, Laura Hornby, Sonny Dhanani, and Jennifer Chandler, "Family Experiences with Non-Therapeutic Research on Dying Patients in the Intensive Care Unit," *Journal of Medical Ethics* 48, no. 11 (2022): 845-851.

203    **they will find useful:** In this it may be aided by thoughtful analyses like the one presented in Nicholas Murphy, Charles Weijer, Derek Debicki, Geoffrey Laforge, Loretta Norton, Teneille Gofton, and Marat Slessarev, "Ethics of Non-Therapeutic Research on Imminently Dying Patients in the Intensive Care Unit," *Journal of Medical Ethics* (2022), epub ahead of print, doi:10.1136/medethics-2021-107953.

204    **OPRR refused to permit:** Zachary M. Schrag, *Ethical Imperialism: Institutional Review Boards and the Social Sciences, 1965-2009* (Baltimore: Johns Hopkins University Press, 2010), 43.

# Works Cited

Advisory Committee on Human Radiation Experiments. *Final Report of the Advisory Committee on Human Radiation Experiments*. New York: Oxford University Press, 1996.

Ågård, A., G. Hermeren, and Johan Herlitz. "Patients' Experiences of Intervention Trials on the Treatment of Myocardial Infarction: Is it Time to Adjust the Informed Consent Procedure to the Patient's Capacity?" *Heart* 86, no. 6 (2001): 632-637.

Allegaert Karel. "Pediatric clinical pharmacology: an introduction to a series of educational papers." *European Journal of Pediatrics* 172:289-292 (2013). doi:10.1007/s00431-012-1921-3.

Altman, Lawrence K. "A Kennedy Baby's Life and Death." *New York Times,* July 29, 2013.

Amdur, Robert J. and Elizabeth A. Bankert. *Institutional Review Board: Member Handbook*, 3rd ed. Sudbury, MA.: Jones and Bartlett, 2011.

American Association of University Professors. "Institutional Review Boards and Social Science Research." 2000. Accessed October 19, 2022, https://www.aaup.org/report/institutional-review-boards-and-social-science-research.

Anderson, Emily E., Stephanie Solomon, Elizabeth Heitman, James M. DuBois, Celia B. Fisher, Rhonda G. Kost, Mary Ellen Lawless, Cornelia Ramsey, Bonnie Jones, and Alice Ammerman. "Research Ethics Education for Community-Engaged Research: A Review and Research Agenda." *Journal of Empirical Research on Human Research Ethics* 7, no. 2 (2012): 3-19.

Anderson, James R., Toby L. Schonfeld, Timothy K. Kelso, and Ernest D. Prentice. "Women in Early Phase Trials: An IRB's Deliberations." *IRB: Ethics & Human Research* 25, no. 4 (2003): 7-11.

Annas, George J. "Old and Emerging Bioethical Issues in Research on Atoms and Genes." *Protecting Human Subjects, U.S. Department of Energy, Office of Biological and Environmental Research* no. Winter (1996): 10.

Annas, George J. "Reforming Informed Consent to Genetic Research." *JAMA: The Journal of the American Medical Association* 286, no. 18 (2001): 2326-2328.

Annas, George J. and Catherine L. Annas. "'Unusual Care': Groupthink and Willful Blindness in the SUPPORT Study." *The American Journal of Bioethics* 20, no. 1 (2020): 44-46.

Annas, George J. and Michael A. Grodin. *The Nazi Doctors and the Nuremberg Code: Human Rights in Human Experimentation*. New York: Oxford University Press, 1992.

Appelbaum, Paul S., Loren H. Roth, and Charles Lidz. "The Therapeutic Misconception: Informed Consent in Psychiatric Research." *International Journal of Law and Psychiatry* 5, no. 3 (1982): 319-329. doi:10.1016/0160-2527(82)90026-7.

ARDS Network. "Ventilation with Lower Tidal Volume as Compared with Traditional Volumes for Acute Lung Injury and the Acute Respiratory Distress Syndrome." *New England Journal of Medicine* 342, (2000): 1301-1308.

Arora, Sameer, George A. Stouffer, Anna M. Kucharska-Newton, Arman Qamar, Muthiah Vaduganathan, Ambarish Pandey, Deborah Porterfield, Ron Blankstein, Wayne D. Rosamond, and Deepak L. Bhatt. "Twenty Year Trends and Sex Differences in Young Adults Hospitalized with Acute Myocardial Infarction: The ARIC Community Surveillance Study." *Circulation* 139, no. 8 (2019): 1047-1056.

Atanasov, Pavel D. "Risk Preferences in Choices for Self and Others: Meta Analysis and Research Directions." *Social Science Research Network* (posted 9/25/2010, last revised 6/17/2015). doi:10.2139/ssrn.1682569.

Aubrey, Harold L. "African Americans and the Broader Legacy of Experience with the American Health Care Community: Parasites, Locusts and Scavengers." In Ralph V. Katz and Rueben C. Warren, eds. *The Search for the Legacy of the USPHS Syphilis Study at Tuskegee*. Lanham, MD: Lexington Books, 2011, 107-115.

Babb, Sarah L. "The Privatization of Human Research Ethics: An American Story." *European Journal for the History of Medicine and Health* 78, no. 2 (2021): 392-411.

Babb, Sarah L. *Regulating Human Research: IRBs from Peer Review to Compliance Bureaucracy*. Stanford, California: Stanford University Press, 2020.

Banner, William. " 'It is Easier to Denature Plutonium than to Denature the Evil Spirit of Man'—Albert Einstein." *Clinical Toxicology* 50, no. 7 (2012): 537.

Baron, Jonathan. "Some Fallacies of Human-Subjects Protection, and some Solutions." *Cortex; a Journal Devoted to the Study of the Nervous System and Behavior* 65, (2014): 246-254.

Baron, Jonathan. *Against Bioethics*. Cambridge, MA: MIT Press, 2006.

Beauchamp, Tom L. "The Origins and Evolution of the Belmont Report." In James F. Childress, Eric M. Meslin, and Harold T. Shapiro, eds, *Belmont Revisited: Ethical Principles for Research with Human Subjects*. Washington, DC: Georgetown University Press, 2005, 12-25.

Becker, Howard S. "Comment on Kevin D. Haggerty, 'Ethics Creep: Governing Social Science Research in the Name of Ethics'." *Qualitative Sociology* 27, no. 4 (2004): 415-416.

Beecher, Henry K. "Ethics and Clinical Research." *The New England Journal of Medicine* 274, no. 24 (June 16, 1966): 1354-1360. doi:10.1056/NEJM196606162742405.

Beecher, Henry K. "Experimentation in Man." *Journal of the American Medical Association* 169, no. 5 (1959): 461-478.

Belluck, Pam. "32 Days on a Ventilator: One Covid Patient's Fight to Breathe Again." *New York Times* 4/26/20, updated 5/8/20.

Ben-Shahar, Omri and Carl E. Schneider. *More than You Wanted to Know: The Failure of Mandated Disclosure.* Princeton: Princeton University Press, 2014.

Berenholtz, Sean, Peter Pronovost, Pamela Lipsett, Deborah Hobson, Karen Earsing, Jason Farley, Shelley Milanovich, et al. "Eliminating Catheter-Related Bloodstream Infections in the Intensive Care Unit." *Critical Care Medicine* 32, no. 10 (Oct, 2004): 2014-2020. doi:10.1097/01.CCM.0000142399.70913.2F.

Bernstein, Peter L. *Against the Gods: The Remarkable Story of Risk.* New York: John Wiley & Sons, 1996.

Best, M. and D. Neuhauser. "Henry K Beecher: Pain, Belief and Truth at the Bedside. the Powerful Placebo, Ethical Research and Anaesthesia Safety." *Quality & Safety in Health Care* 19, no. 5 (Oct, 2010): 466-468. doi:10.1136/qshc.2010.042200.

Bledsoe, Caroline H., Bruce Sherin, Adam G. Galinsky, Nathalia M. Headley, Carol A. Heimer, Erik Kjeldgaard, James Lindgren, Jon D. Miller, Michael E. Roloff, and David H. Uttal. "Regulating Creativity: Research and Survival in the IRB Iron Cage." *Northwestern University Law Review* 101, no. 2 (Feb 1, 2007): 593-641.

Blumgart, Herrman L. "The Medical Framework for Viewing the Problem of Human Experimentation." *Daedalus* (1969): 248-274.

Borror, Kristina C., Director, Division of Compliance Oversight, OHRP. *Letter to Daniel E. Ford, Vice Dean for Clinical Investigation, Johns Hopkins, July 19, 2007,* re ICU infection study.

Brainard, Jeffrey and D. W. Miller. "Spate of Suspensions of Academic Research Spurs Questions about Federal Strategy." *The Chronicle of Higher Education* 46, no. 22 (Feb 4, 2000): A29.

Brecher, Ruth and Edward Brecher. "They Volunteered for Cancer." *Reader's Digest,* April, 1958, 62-66.

Breyer, Stephen. *Breaking the Vicious Circle: Toward Effective Risk Regulation.* Cambridge, MA: Harvard University Press, 1993.

Buchanan, Lisa R, Division of Compliance Oversight, OHRP. Letter to Richard B. Marchase, V.P. for Research and Development, University of Alabama at Birmingham, March 7, 2013.

Buchanan, Lisa R., Division of Compliance Oversight, OHRP. Letter to Richard B. Marchase, V.P. for Research & Economic Development, University of Alabama at Birmingham, June 4, 2013.

Burnett, Charles H., Esther L. Bloomberg, Gerald Shortz, David W. Compton, and Henry K. Beecher. "A Comparison of the Effects of Ether and Cyclopropane Anesthesia on the Renal Function of Man." *Journal of Pharmacology and Experimental Therapeutics* 96, no. 4: 380-387.

Burris, Scott. "Regulatory Innovation in the Governance of Human Subjects Research: A Cautionary Tale and some Modest Proposals." *Regulation & Governance* 2, no. 1 (Mar, 2008): 65-84. doi:10.1111/j.1748-5991.2007.00025.x.

Bush, Lawrence. "July 29: The Tuskegee Syphilis Experiment." *Jewish Currents,* July 28, 2015.

Buxton, Peter. "Testimony by Peter Buxton from the United States Senate Hearings on Human Experimentation." In *Tuskegee's Truths: Rethinking the Tuskegee Syphilis Study*, ed. Susan Reverby. Chapel Hill and London: University of North Carolina Press, 2000, pp. 150-156. His name is usually spelled Buxtun, but in the Hearings it is reported as Buxton.

Byrd, W. Michael and Linda A. Clayton. *An American Health Dilemma, Volume I: A Medical History of African Americans and the Problem of Race, Beginnings to 1900.* New York: Routledge, 2000.

Byrd, W. Michael and Linda A. Clayton. *An American Health Dilemma, Volume II: Race, Medicine, and Health Care in the United States 1900–2000.* New York: Routledge, 2002.

Candilis, Philip J., Rasim Arikan, Sheila B. Noone, and Jacob C. Holzer. "The New Research Ethic: Will Oversight Requirements Sink Forensic Research?" *The Journal of the American Academy of Psychiatry and the Law* 33, no. 3 (2005): 361.

Capps, Richard B., Alfred M. Bennett, and Joseph Stokes. "Endemic Infectious Hepatitis in an Infants' Orphanage: I. Epidemiologic Studies in Student Nurses." *AMA Archives of Internal Medicine* 89, no. 1 (1952): 6-23.

Capron, Alexander M. "The Dog in the Night-Time: Or, the Curious Relationship of the Belmont Report and the President's Commission." In *Belmont Revisited: Ethical Principles for*

*Research with Human Subjects*, ed. by James F. Childress, Eric M. Meslin, and Harold T. Shapiro. Washington, DC: Georgetown University Press, 2005, 29-40.

Carlo, Waldemar A., Edward F. Bell, and Michele C. Walsh. "Oxygen-Saturation Targets in Extremely Preterm Infants." *New England Journal of Medicine* 368, no. 20 (2013): 1949-1950.

Carlo, Wally and Namasivayan Ambalavanan. Consent Form for the Surfactant Positive Airway Pressure and Pulse Oximetry Trial in Extremely Low Birth Weight Infants. 2005. Accessed on October 13, 2022 from NICHHD at https://www.nichd.nih.gov/sites/default/-files/about/Documents/2013-04-08_All_Consent_SUPPORT.pdf.

Carome, Michael A. Public Citizen letter to Secretary Sebelius about the SUPPORT Trial. April 10, 2013.

Carpenter, Dale. "Institutional Review Boards, Regulatory Incentives, and some Modest Proposals for Reform." *Northwestern University Law Review* 101, no. 2 (Feb 1, 2007): 687.

Cassileth, Barrie R., Robert V. Zupkis, Katherine Sutton-Smith, and Vicki March. "Informed Consent—Why are its Goals Imperfectly Realized?" *The New England Journal of Medicine* 302, no. 16 (1980): 896.

Ceci, Stephen J., Douglas Peters, and Jonathan Plotkin. "Human Subjects Review, Personal Values, and the Regulation of Social Science Research." *The American Psychologist* 40, no. 9 (1985): 994-1002. doi:10.1037//0003-066X.40.9.994.

Centers for Disease Control and Prevention. "U.S. Public Health Service Syphilis Study at Tuskegee: Research Implications—How Tuskegee Changed Research Practices." Updated 2020. Accessed April 28, 2020, https://www.cdc.gov/tuskegee/after.htm.

Christie, D. R. H., G. S. Gabriel, and K. Dear. "Adverse Effects of a Multicentre System for Ethics Approval on the Progress of a Prospective Multicentre Trial of Cancer Treatment: How Many Patients Die Waiting?" *Internal Medicine Journal* 37, no. 10 (Oct, 2007): 680-686. doi:10.1111/j.1445-5994.2007.01451.x.

Coe, Fredric L. "The Costs and Benefits of a Well-Intended Parasite: A Witness and Reporter on the IRB Phenomenon." *Northwestern University Law Review* 101, no. 2 (Feb 1, 2007): 723-733.

Coe, Fredric L. and Lynn Raisen. "Allopurinol Treatment of Uric-Acid Disorders in Calcium-Stone Formers." *Lancet* 301, no. 7795 (1973), 129-131.

Cohen, I. Glenn. "Introduction to Part V—Paradigm Shifts in Research Ethics." In I. Glenn Cohen and Holly Fernandez Lynch, eds, *Human Subjects Research Regulation: Perspectives on the Future.* Cambridge, MA: MIT Press, 2014, 281-284.

Cole, C. H., K. W. Wright, W. Tarnow-Mordi, D.L. Phelps, and the Pulse Oximetry Saturation Trial for Prevention of Retinopathy of Prematurity Planning Study Group. "Resolving our Uncertainty about Oxygen Therapy." *Pediatrics* 112, no. 6 Pt 1 (Dec, 2003): 1415-1419. doi:10.1542/peds.112.6.1415.

Collins, Rory, Richard Doll, and Richard Peto. "Ethics of Clinical Trials." In C. J. Williams, ed., *Introducing New Treatments for Cancer: Practical, Ethical, and Legal Problems.* Chichester; New York: Wiley, 1992, 49-65.

Committee of Sponsoring Organizations of the Treadway Commission (COSO). *Enterprise Risk Management—Integrating with Strategy and Performance, Executive Summary.* 2017. Accessed on October 18, 2022, https://www.coso.org/sitepages/guidance-on-enter-prise-risk-management.aspx?web=1,

Corbie-Smith, Giselle. "The Continuing Legacy of the Tuskegee Syphilis Study: Considerations for Clinical Investigation." *The American Journal of the Medical Sciences* 317, no. 1 (1999): 5-8.

Cowan, D. H. "Human Experimentation: The Review Process in Practice." *Case Western Reserve Law Review* 25, no. 3 (1975): 533-564.

Curran, William J. "Governmental Regulation of the use of Human Subjects in Medical Research: The Approach of Two Federal Agencies." *Daedalus* 98, no. 2 (1969): 542-594.

Dance, A. "Survival of the Littlest: The Long-Term Impacts of being Born Extremely Early." *Nature* 582, no. 7810 (Jun, 2020): 20-23. doi:10.1038/d41586-020-01517-z.

Daugherty, Christopher K., Donald M. Banik, Linda Janish, and Mark J. Ratain. "Quantitative Analysis of Ethical Issues in Phase I Trials: A Survey Interview Study of 144 Advanced Cancer Patients." *IRB: Ethics & Human Research* 22, no. 3 (2000): 6-14.

Department of Health and Human Services. "Public Meeting: Matters Related to Protection of Human Subjects and Research Considering Standard of Care Interventions." August 28, 2013, https://www.-hhs.gov/ohrp/regulations-and-policy/requests-for-comments/-public-Meeting-08-28-2013/index.Html. Accessed February 14, 2021.

Department of Health, Education, and Welfare. *Final Report of the Tuskegee Syphilis Study Ad Hoc Advisory Panel.* Washington, DC: Government Printing Office, 1973.

Department of Homeland Security and 15 other departments and agencies. *Final Rule, 82 Fed Reg 7149-7273.* January 19, 2017.

Department of Homeland Security and 15 other departments and agencies. *Federal Policy for the Protection of Human Subjects: Delay of the Revisions to the Federal Policy for the Protection of Human Subjects. 83 Fed Reg 2885-94.* January 22, 2018.

Dickert, Neal W., Alex Llanos, and Habib Samady. "Re-Visiting Consent for Clinical Research on Acute Myocardial Infarction and Other Emergent Conditions." *Progress in Cardiovascular Diseases* 55, no. 3 (2012): 251-257.

Dickert, Neal W., K. A. Hendershot, C. D. Speight, and A. E. Fehr. "Patients' Views of Consent in Clinical Trials for Acute Myocardial Infarction: Impact of Trial Design." *Journal of Medical Ethics* 43, no. 8 (Aug, 2017): 524-529. doi:10.1136/medethics-2016-103866.

Dingwall, Robert. "The Ethical Case Against Ethical Regulation in Humanities and Social Science Research." *Twenty-First Century Society* 3, no. 1 (2008): 1-12.

Dingwall, Robert. "Turn Off the Oxygen..." *Law & Society Review* 41, no. 4 (2007): 787-795.

Doherty, Neil A. *Integrated Risk Management Techniques and Strategies for Managing Corporate Risk.* New York: McGraw-Hill, 2000.

Dresser, Rebecca and Joel Frader. "Off-Label Prescribing: A Call for Heightened Professional and Government Oversight." *Journal of Law, Medicine & Ethics* 37, no. 3 (2009): 476-486.

Dresser, Rebecca. *When Science Offers Salvation: Patient Advocacy and Research Ethics.* Oxford; New York: Oxford University Press, 2001.

Echt, Debra S., Philip R. Liebson, L. Brent Mitchell, Robert W. Peters, Dulce Obias-Manno, Allan H. Barker, Daniel Arensberg, Andrea Baker, Lawrence Friedman, and H. Leon Greene. "Mortality and Morbidity in Patients Receiving Encainide, Flecainide, Or Placebo: The Cardiac Arrhythmia Suppression Trial." *New England Journal of Medicine* 324, no. 12 (1991): 781-788.

El-Hinnawy, Patricia C. "Interview with Norman Fost" in *OHRP: Oral History of the Belmont Report and the National Commission for the Protection of Human Subjects of Biomedical and Behavioral Research.* Belmont Oral History Project, 2004.

Elders, M. Joycelyn. "Foreword: The Search for the Legacy of the USPHS Syphilis Study at Tuskegee." In Ralph V. Katz and Rueben C. Warren, eds. *The Search for the Legacy of the USPHS Syphilis Study at Tuskegee.* Lanham, MD: Lexington Books, 2011, ix-xiv.

Elliott, Carl. "The Tyranny of Expertise." In *The Ethics of Bioethics: Mapping the Moral Landscape,* ed. Lisa A. Eckenwiler and Felicia G. Cohn. Baltimore: Johns Hopkins University Press, 2007. 43-47.

Emanuel, Ezekiel J., Anne Wood, Carianne Tucker Muse, Jeremy Sugarman, Alan Fleischman, Angela Bowen, Kenneth A. Getz, et al. "Oversight of Human Participants Research: Identifying Problems to Evaluate Reform Proposals." *Annals of Internal Medicine* 141, no. 4 (Aug 17, 2004): 282-291. doi:10.7326/0003-4819-141-4-200408170-00008.

Epstein, Lynn Chaikin and Louis Lasagna. "Obtaining Informed Consent: Form Or Substance." *Archives of Internal Medicine* 123, no. 6 (1969): 682-688.

Evans, Barbara J. "The First Amendment Right to Speak about the Human Genome." *University of Pennsylvania Journal of Constitutional Law* 16, no. 3 (2014): 549.

Evans, Barbara J. "The Genetic Information Nondiscrimination Act at Age 10: GINA's Controversial Assertion that Data Transparency Protects Privacy and Civil Rights." *William and Mary Law Review* 60, no. 6 (2019): 2017.

Eyal, Nir and Tobias Gerhard. "Do Coronavirus Vaccine Challenge Trials have a Distinctive Generalisability Problem?" *Journal of Medical Ethics* 48, no. 9 (2022): 586-589.

Eyal, Nir, Marc Lipsitch, and Peter G. Smith. "Human Challenge Studies to Accelerate Coronavirus Vaccine Licensure." *The Journal of Infectious Diseases* 221, no. 11 (2020): 1752-1756.

Eyal, Nir. "Is There an Ethical Upper Limit on Risks to Study Participants?" *Public Health Ethics* 13:2 (2020): 143-156.

Faden, Ruth R., Nancy E. Kass, Steven N. Goodman, Peter Pronovost, Sean Tunis, and Tom L. Beauchamp. "An Ethics Framework for a Learning Health Care System: A Departure from Traditional Research Ethics and Clinical Ethics." *The Hastings Center Report* 43, no. s1 (Jan, 2013): S16-S27. doi:10.1002/hast.134.

Faden, Ruth R., Tom L. Beauchamp, and Nancy M. P. King. *A History and Theory of Informed Consent.* New York: Oxford University Press, 1986.

Fairchild, Amy L. and Ronald Bayer. "Uses and Abuses of Tuskegee." *Science* 284, no. 5416 (1999): 919-921.

Fitzgerald, Maureen H. "Big Basket Or Mission Creep?" *Professional Ethics Report* XVII, no. 2 (2004): 1.

Fitzgerald, Maureen H. "Punctuated Equilibrium, Moral Panics and the Ethics Review Process." *Journal of Academic Ethics* 2, no. 4 (2005): 315-338.

Fitzgerald, Maureen H. and Paul A. Phillips. "Centralized and Non-Centralized Ethics Review: A Five Nation Study." *Accountability in Research* 13, no. 1 (2006): 47-74.

Fitzgerald, Maureen H., Paul A. Phillips, and Elisa Yule. "The Research Ethics Review Process and Ethics Review Narratives." *Ethics & Behavior* 16, no. 4 (2006): 377-395.

Flory, James and Ezekiel Emanuel. "Interventions to Improve Research Participants' Understanding in Informed Consent for Research: A Systematic Review." *JAMA: The Journal of the American Medical Association* 292, no. 13 (2004): 1593-1601.

Fost, Norman and Robert J. Levine. "The Dysregulation of Human Subjects Research." *JAMA: The Journal of the American Medical Association* 298, no. 18 (Nov 14, 2007): 2196-2198.

Fost, Norman. "Ethical Dilemmas in Medical Innovation and Research: Distinguishing Experimentation from Practice." *Seminars in Perinatology* 22, no. 3 (June 1998): 223-232.

Fost, Norman. "Fusing and Confusing Ethics and Regulations." In Paula Knudson, Bridget Gardner and Joan Rachlin, eds., *PRIM&R Through the Years: Three Decades of Protecting Human Subjects*. Boston, MA: Public Responsibility in Medicine and Research, 2006, 454-464

Frankel, Mark S. "The Public Health Service Guidelines Governing Research Involving Human Subjects: An Analysis of the Policy-Making Process." PhD dissertation, George Washington University, 1972.

Freidenfelds, Lara. "Recruiting Allies for Reform: Henry Knowles Beecher's "Ethics and Clinical Research"." *International Anesthesiology Clinics* 45, no. 4 (2007): 79-103.

Fried, Charles. *Medical Experimentation: Personal Integrity and Social Policy*. Amsterdam; New York: North-Holland Pub. Co.; American Elsevier, 1974.

Friends Service Alliance. "Compliance and Risk Management Partners." N.d. Accessed on November 15, 2022, https://www.fsainfo.org/-compliance-partners.

Furman, Lydia M. "Rebuttal to the Python's Embrace." *Pediatrics* 129:3, March 2012.

Gammelgaard, Anne, Peter Rossel, and Ole Steen Mortensen. "Patients' Perceptions of Informed Consent in Acute Myocardial Infarction Research: A Danish Study." *Social Science & Medicine* 58, no. 11 (2004): 2313-2324.

Gawande, Atul. "A Lifesaving Checklist" (Op-Ed). *New York Times*, December 30, 2007.

General Accounting Office. *Scientific Research: Continued Vigilance Critical to Protecting Human Subjects: Report to the Ranking Minority Member, Committee on Governmental Affairs, U.S. Senate*. District of Columbia: General Accounting Office, 1996. GAO/HEHS-96-72.

Gilbert, Jack, Rob Knight, and Sandra Blakeslee. *Dirt is Good: The Advantage of Germs for Your Child's Developing Immune System*. New York, NY: St. Martin's Press, 2017.

Goldfarb, Norman M. "Greg Koski on Human Subjects Protection." *Journal of ClinicalResearch and Best Practices* 3, no. 9 (2007): 1-6.

Goodman, Walter. "Doctors must Experiment on Humans but What are the Patient's Rights?" *New York Times Magazine* July 2, 1967.

Gordon, Bruce G., Joseph Brown, Christopher Kratochvil, Ernest D. Prentice, and Toby L. Schonfeld. "Paying Research Subjects." In *Institutional Review Board Management and Function*, 2nd ed, ed. Elizabeth A. Bankert and Robert J. Amdur. Burlington, MA: Jones & Bartlett Learning, 2006, 154-159.

Gravenstein, J. S. "Henry K. Beecher: The Introduction of Anesthesia into the University." *Anesthesiology* 88, no. 1 (1998): 245-253.

Gray, Fred D. *The Tuskegee Syphilis Study: The Real Story and Beyond.* Montgomery, AL: NewSouth Books, 1998.

Greely, Henry T. "Legal, Ethical, and Social Issues in Human Genome Research." *Annual Review of Anthropology* (1998): 473-502.

Greely, Henry T. and Mildred K. Cho. "The Henrietta Lacks Legacy Grows." *EMBO Reports; EMBO Rep* 14, no. 10 (2013): 849. doi:10.1038/embor.2013.148.

Gunsalus, C. Kristina, Edward M. Bruner, Nicholas C. Burbules, Leon Dash, Matthew Finkin, Joseph P. Goldberg, William T. Greenough, et al. "The Illinois White Paper: Improving the System for Protecting Human Subjects: Counteracting IRB 'Mission Creep'." *Qualitative Inquiry* 13, no. 5 (Jul, 2007): 617-649. doi:10.1177/1077800407300785.

Halpern, Sydney A. *Lesser Harms: The Morality of Risk in Medical Research.* Chicago: University of Chicago Press, 2004.

Halpern, Sydney. "Hybrid Design, Systemic Rigidity: Institutional Dynamics in Human Research Oversight." *Regulation & Governance* 2, no. 1 (Mar, 2008): 85-102. doi:10.1111/j.1748-5991.2007.00032.x.

Hampton, John. "Commentary: The Need for Clinical Freedom." *International Journal of Epidemiology* 40, no. 4 (2011): 849-852.

Hansson, Mats. "Genomics and society—Ethical, Legal, Cultural and Socioeconomic Implication." *European Journal of Human Genetics* 24, no. 12 (2016): 1830.

Heaton, Leonard D., John B. Coates Jr., Ebbe C. Hoff and Phebe M. Hoff. "Preventive Medicine in World War II. Volume 5. Communicable Diseases Transmitted through Contact Or by Unknown Means," Office of the Surgeon General (Army), 1960.

Heimer, Carol A. "The Unstable Alliance of Law and Morality." In Steven Hitlin and Stephen Vaisey, eds., *Handbook of the Sociology of Morality.* New York, NY: Springer, 2010, 179-202.

Hopkin, Paul. *Fundamentals of Risk Management,* 4th ed. New York, NY: Kogan Page, 2017.

House of Representatives, Subcommittee on Human Resources, Committee on Government Reform and Oversight (1998). *Institutional Review Boards: A System in Jeopardy.* 105th Congress, Second Session, Serial No. 105-166.

Hudson, Kathy L., Alan E. Guttmacher, and Francis S. Collins. "In Support of SUPPORT—a View from the NIH." *New England Journal of Medicine* 368, no. 25 (2013): 2349-2351.

Hunter, David. "Is Research Ethics Regulation Really Killing People?" *Medical Journal of Australia* 202, no. 6 (2015): 338-339.

Hyman, David A. "Institutional Review Boards: Is this the Least Worst we can do?" *Northwestern University Law Review* 101, no. 2 (Feb 1, 2007): 749.

Hyman, David A. "The Pathologies of Institutional Review Boards: Are IRBs the 'Least Worst' Way to Promote Research Ethics?" *Regulation* 30, (2007): 42-49.

Institute of Medicine. *Beyond the HIPAA Privacy Rule: Enhancing Privacy, Improving Health Through Research.* Washington, DC: The National Academies Press, 2009.

Institute of Medicine. *Health Standards for Long Duration and Exploration Spaceflight: Ethics Principles, Responsibilities, and Decision Framework.* Washington, DC: The National Academies Press, 2014, https://doi.org/10.17226/18576.

ISIS-2 (Second International Study of Infarct Survival) Collaborative Group. "Randomised Trial of Intravenous Streptokinase, Oral Aspirin, both, or Neither among 17,187 Cases of Suspected Acute Myocardial Infarction: ISIS-2." *Lancet* 332, (1988): 349-360.

Jacobs, Noortje. *Ethics by Committee: A History of Reasoning Together about Medicine, Science, Society, and the State.* Chicago, IL: University of Chicago Press, 2022.

Jacobson, Sherry. "Dallas Researchers Fail to Explain Why Parents in National Baby Study Weren't Notified of Fatal Risks." *Dallas Morning News,* April 11, 2013.

Jacobson, Sherry. "UTSW Enrolled 73 Premature Infants in Controversial Study; 17 of them Died." *Dallas Morning News,* April 13, 2013.

Joffe, Steven. "Revolution Or Reform in Human Subjects Research Oversight." *Journal of Law, Medicine & Ethics* 40, no. 4 (2012): 922-929.

Jonas, Hans. "Philosophical Reflections on Experimenting with Human Subjects." *Daedalus* 98, no. 2 (1969): 219-247, https://www.jstor.org/stable/20023877.

Jones, David S., Christine Grady, and Susan E. Lederer. "'Ethics and Clinical Research'— the 50th Anniversary of Beecher's Bombshell." *The New England Journal of Medicine* 374, no. 24 (2016): 2393-2398. doi:10.1056/NEJMms1603756.

Jones, James H. *Bad Blood: The Tuskegee Syphilis Experiment.* New York and London: Free Press, 1981.

Jonsen, Albert R. "On the Origins and Future of the Belmont Report." In James F. Childress, Eric M. Meslin, and Harold T. Shapiro, eds, *Belmont Revisited: Ethical Principles for Research with Human Subjects.* Washington, DC: Georgetown University Press, 2005, 3-11.

Kahn, Jeffrey P., Leslie Meltzer Henry, Anna C. Mastroianni, Wilbur H. Chen, and Ruth Macklin. "For Now, It's Unethical to use Human Challenge Studies for SARS-CoV-2 Vaccine Development." *Proceedings of the National Academy of Sciences* 117, no. 46 (2020): 28538-28542.

Kallet, Richard H. "What is the Legacy of the National Institutes of Health Acute Respiratory Distress Syndrome Network?" *Respiratory Care* 54, no. 7 (2009): 912-924.

Kapadia, V. S., L. F. Chalak, J. E. Sparks, J. R. Allen, R. C. Savani, and M. H. Wyckoff. "Resuscitation of Preterm Neonates with Limited Versus High Oxygen Strategy." *Pediatrics* 132, no. 6 (Dec, 2013): e1488-96. doi:10.1542/peds.2013-0978.

Kass, Nancy, Ruth Faden, Rachel E. Fabi, Stephanie Morain, Kristina Hallez, Danielle Whicher, Sean Tunis, Rachael Moloney, Donna Messner, and James Pitcavage. "Alternative Consent Models for Comparative Effectiveness Studies: Views of Patients from Two Institutions." *AJOB Empirical Bioethics* 7, no. 2 (2016): 92-105.

Katz, Jay, Alexander Morgan Capron, and Eleanor Swift Glass. *Experimentation with Human Beings: the Authority of the Investigator, Subject, Professions, and State in the Human Experimentation Process.* New York: Russell Sage Foundation, 1972.

Katz, Ralph V. and Rueben C. Warren, eds. *The Search for the Legacy of the USPHS Syphilis Study at Tuskegee.* Lanham, MD: Lexington Books, 2011.

Kennedy, Thomas J., Jr. "James Augustine Shannon." In *Biographical Memoirs.* Washington, DC: National Academies Press, 1994.

King, Nancy M. P. "Key Information in the New Common Rule: Can it Save Research Consent?" *Journal of Law, Medicine & Ethics* 47, no. 2 (2019): 203-212.

King, Nancy M. P., Gail Henderson and Jane Stein. *Beyond Regulations: Ethics in Human Subjects Research.* Chapel Hill, NC: University of North Carolina Press, 1999.

King, Patricia A. "Justice Beyond *Belmont.*" In *Belmont Revisited: Ethical Principles for Research with Human Subjects,* ed. James F. Childress, Eric M. Meslin and Harold T. Shapiro. Washington, DC: Georgetown University Press, 2005, 136-147.

Klitzman, Robert L. "Local IRBs Vs. Federal Agencies: Shifting Dynamics, Systems, and Relationships." *Journal of Empirical Research on Human Research Ethics: JERHRE* 7, no. 3 (Jul, 2012): 50-62. doi:10.1525/jer.2012.7.3.50.

Klitzman, Robert L. *The Ethics Police?: The Struggle to make Human Research Safe.* Oxford; New York: Oxford University Press, 2015.

Kopp, Vincent J. "Henry Knowles Beecher and the Development of Informed Consent in Anesthesia Research." *Anesthesiology: The Journal of the American Society of Anesthesiologists* 90, no. 6 (1999): 1756-1765.

Koski, Greg. "Beyond Compliance...is it Too Much to Ask?" *IRB: Ethics & Human Research* 25, no. 5 (2003): 5-6.

Koski, Greg. "Changing the Paradigm: New Directions in Federal Oversight of Human Research." *Journal of Pediatric Gastroenterology and Nutrition* 37 Suppl 1, no. Supplement 1 (Nov, 2003): S2-S6. doi:10.1097/00005176-200311001-00002.

Koski, Greg. "Getting Past Protectionism: Is it Time to Take Off the Training Wheels?" In I. Glenn Cohen and Holly Fernandez Lynch, eds, *Human Subjects Research Regulation: Perspectives on the Future*. Cambridge, MA: MIT Press, 2014.

Koski, Greg. "Research Ethics and Oversight: Revolution, Or just Going Around in Circles?" *The Monitor* 21, no. 55 (2007): 2007-2057.

Krugman, Saul, Robert Ward, Joan P. Giles, Oscar Bodansky, and A. Milton Jacobs. "Infectious Hepatitis: Detection of Virus during the Incubation Period and in Clinically Inapparent Infection." *New England Journal of Medicine* 261, no. 15 (1959): 729-734.

Krugman, Saul. "The Willowbrook Hepatitis Studies Revisited: Ethical Aspects." *Reviews of Infectious Diseases* 8, no. 1 (1986): 157-162.

Kuehn, Bridget M. "DHHS Halts Quality Improvement Study." *JAMA: The Journal of the American Medical Association* 299, no. 9 (Mar 5, 2008): 1005-1006. doi:10.1001/jama.299.9.1005.

Kulynych, Jennifer and Henry T. Greely. "Clinical Genomics, Big Data, and Electronic Medical Records: Reconciling Patient Rights with Research when Privacy and Science Collide." *Journal of Law and the Biosciences* 4, no. 1 (2017): 94-132.

Ladimer, Irving and Roger W. Newman. *Clinical Investigation in Medicine: Legal, Ethical, and Moral Aspects: an Anthology and Bibliography*. Boston: Law-Medicine Research Institute, Boston University, 1963.

Lancaster, John and James P. Rathwell. "The Moralist." *The Medium*, July 14, 2016. Accessed October 16, 2022, https://medium.com-/thebigroundtable/the-moralist-ad8159-ebe6be.

Lantos, John D. "Learning the Right Lessons from the SUPPORT Study Controversy." *Archives of Disease in Childhood, Fetal and Neonatal Edition* 99, no. 1 (Jan, 2014): F4-5. doi:10.1136/archdischild-2013-304916 .

Lantos, John D. "OHRP and Public Citizen are Wrong about Neonatal Research on Oxygen Therapy." *Bioethics Forum* 2020, no. 11/16 (2013).

Lantos, John D. "The 'Inclusion Benefit' in Clinical Trials." *The Journal of Pediatrics* 134, no. 2 (Feb, 1999): 130-131. doi:S0022-3476(99)70400-2 [pii].

Lantos, John D. "The Weird Divergence of Ethics and Regulation with Regard to Informed Consent." *The American Journal of Bioethics* 13, no. 12 (2013): 31-33.

Lantos, John D. "Vindication for SUPPORT." *New England Journal of Medicine* 373, no. 15 (2015): 1393-1395.

Lasagna, Louis. "Special Subjects in Human Experimentation." *Daedalus* 98, no. 2 (1969): 449-462.

Lebacqz, Karen. "We Sure are Older but are We Wiser?" In James F. Childress, Eric M. Meslin, and Harold T. Shapiro, eds, *Belmont Revisited: Ethical Principles for Research with Human Subjects.* Washington, DC: Georgetown University Press, 2005, 99-110.

Lederer Susan E. *Subjected to Science: Human Experimentation in America Before the Second World War.* Baltimore: Johns Hopkins University Press, 1995.

Levine, Robert J. *Ethics and Regulation of Clinical Research*, 2nd ed. Baltimore: Urban & Schwarzenberg, 1986.

Levy, David J. *Hans Jonas: The Integrity of Thinking.* Vol. 1. Columbia, MO: University of Missouri Press, 2002.

Lynch, Holly Fernandez. "Protecting Human Research Subjects as Human Research Workers." In I. Glenn Cohen and Holly Fernandez Lynch, eds, *Human Subjects Research Regulation: Perspectives on the Future.* Cambridge, MA: MIT Press, 2014, 327-340.

Macklin, Ruth, Lois Shepherd, Alice Dreger, Adrienne Asch, Francoise Baylis, Howard Brody, Larry R. Churchill, Carl H. Coleman, Ethan Cowan, and Janet Dolgin. "The OHRP and SUPPORT—Another View." *New England Journal of Medicine* 369, no. 2 (2013): e3.

Malone, Ruth E., Valerie B. Yerger, Carol McGruder, and Erika Froelicher. " 'It's Like Tuskegee in Reverse': A Case Study of Ethical Tensions in Institutional Review Board Review of Community-Based Participatory Research." *American Journal of Public Health* 96, no. 11 (2006): 1914-1919.

Marshall, Eliot. "Shutdown of Research at Duke Sends a Message." *Science* 284, no. 5418 (1999): 1246.

Mashour, George A. "Altered States: LSD and the Anesthesia Laboratory of Henry Knowles Beecher." *Environment* 69, (2009).

Mashour, George A. "From LSD to the IRB: Henry Beecher's Psychedelic Research and the Foundation of Clinical Ethics." *International Anesthesiology Clinics* 45, no. 4 (2007): 105-111.

Mathews, Jay. "Father's Complaints Shut Down Research." Washington Post, January 12, 2000.

Mazur, Dennis John. *Evaluating the Science and Ethics of Research on Humans: A Guide for IRB Members*. Baltimore: Johns Hopkins University Press, 2007.

McCaffrey, James. "Hospital Accused on Cancer Study." *New York Times* January 21, (1964).

McConnell, G. "The Transplantation of Human Carcinomatous Material into Lower Animals." *The Journal of Experimental Medicine* 10, no. 1 (Jan 1, 1908): 36-44. doi:10.1084/jem.10.1.36.

McDonald, Michael, Susan Cox, and Anne Townsend. "Toward Human Research Protection that is Evidence Based and Participant Centered." In I. Glenn Cohen and Holly Fernandez Lynch, eds, *Human Subjects Research Regulation: Perspectives on the Future*. Cambridge, MA: MIT Press, 2014.

McGuire, Amy L. and Laura M. Beskow. "Informed Consent in Genomics and Genetic Research." *Annual Review of Genomics and Human Genetics* 11, (2010): 361.

Mello, Michelle M., David M. Studdert, and Troyen A. Brennan. "The Rise of Litigation in Human Subjects Research." *Annals of Internal Medicine* 139, no. 1 (2003): 40-45.

Menikoff, Jerry and Edward P. Richards. *What the Doctor Didn't Say: The Hidden Truth about Medical Research*. Oxford: Oxford University Press, 2006.

Menikoff, Jerry. "Where's the Law? Uncovering the Truth about IRBs and Censorship." *Northwestern University Law Review* 101, no. 2 (Feb 1, 2007): 791-799.

Meyer, Michelle N. "Three Challenges for Risk-Based (Research) Regulation: Heterogeneity among Regulated Activities, Regulator Bias, and Stakeholder Heterogeneity." In I. Glenn Cohen and Holly Fernandez Lynch, eds, *Human Subjects Research Regulation: Perspectives on the Future*. Cambridge, MA: MIT Press, 2014.

Miller, Franklin G. and Howard Brody. "Clinical Equipoise and the Incoherence of Research Ethics." *The Journal of Medicine and Philosophy* 32, no. 2 (2007): 151-165.

Miller, Franklin G. and Howard Brody. "What Makes Placebo-Controlled Trials Unethical?" *The American Journal of Bioethics* 2, no. 2 (2002): 3-9.

Miller, Franklin G. and Jonathan Kimmelman. "Introduction to the Special Issue on the Belmont Report." *Perspectives in Biology and Medicine* 63, no. 2 (2020): 219.

Moir, Melinda S., E. Bair, P. Shinnick, and A. Messner. "Acetaminophen Versus Acetaminophen with Codeine After Pediatric Tonsillectomy." *The Laryngoscope* 110, no. 11 (2000): 1824-7.

Moseley, J. Bruce, Kimberly O'Malley, Nancy J. Petersen, Terri J. Menke, Baruch A. Brody, David H. Kuykendall, John C. Hollingsworth, Carol M. Ashton, and Nelda P. Wray. "A Controlled Trial of Arthroscopic Surgery for Osteoarthritis of the Knee." *New England Journal of Medicine* 347, no. 2 (2002): 81-88.

Murphy, Nicholas, Charles Weijer, Derek Debicki, Geoffrey Laforge, Loretta Norton, Teneille Gofton, and Marat Slessarev. "Ethics of Non-Therapeutic Research on Imminently Dying Patients in the Intensive Care Unit." *Journal of Medical Ethics* (2022). doi:10.1136/medethics-2021-107953.

NASA. "Risk Management." Updated September 18, 2022. Accessed November 15, 2022, https://sma.nasa.gov/sma-disciplines/risk-management.

National Commission for the Protection of Human Subjects of Biomedical and Behavioral Research. *The Belmont Report: Ethical Principles and Guidelines for the Protection of Human Subjects of Research.* DHEW Publication no. (OS) 78-0012. Washington, DC: Government Printing Office, 1978.

National Institutes of Health Revitalization Act of 1993, Public Law 103-43, 103d Congress, June 10, 1993.

National Institutes of Health. "NIH Inclusion Outreach Toolkit: How to Engage, Recruit, and Retain Women in Clinical Research." N.d. Accessed April 25, 2020, https://orwh.od.nih.gov/toolkit/-recruitment/history.

National Institutes of Health. "U.S. Clinical Trial Results show Novavax Vaccine is Safe and Prevents COVID-19." July 14, 2021. Accessed August 17, 2021, https://www.nih.gov/news-events/news-releases-/us-clinical-trial-results-show-novavax-vaccine-safe-prevents-covid-19.

Nelson, Daniel K. "Conflict of Interest: Institutional Review Boards." In *Institutional Review Board: Management and Function*, 2nd ed, ed. by Elizabeth A. Bankert and Robert J. Amdur. Burlington, MA: Jones & Bartlett Learning, 2006, 208-212.

Norman Fost and Robert J. Levine, "The Dysregulation of Human Subjects Research." *JAMA: The Journal of the American Medical Association* 298, no. 18 (Nov 14, 2007): 2196-2198. doi:10.1001/jama.298.18.2196.

O'Rourke, P. Pearl. "The Final Rule: When the Rubber Meets the Road." *The American Journal of Bioethics* 17, no. 7 (2017): 27-33.

Oakes, J. Michael. "Risks and Wrongs in Social Science Research." *Evaluation Review* 26, no. 5 (Oct, 2002): 443-479. doi:10.1177/019384102236520.

Office of the Surgeon General, US Public Health Service. "Surgeon General's Directives on Human Experimentation." *American Psychologist* 22, (1967): 350-355.

Offit, Paul A. *Vaccinated: One Man's Quest to Defeat the World's Deadliest Diseases*. New York: Harper Collins, 2008.

Oh, Sam S., Joshua Galanter, Neeta Thakur, Maria Pino-Yanes, Nicolas E. Barcelo, Marquitta J. White, Danielle M. de Bruin, Ruth M. Greenblatt, Kirsten Bibbins-Domingo, and Alan HB Wu. "Diversity in Clinical and Biomedical Research: A Promise Yet to be Fulfilled." *PLoS Medicine* 12, no. 12 (2015): e1001918.

OHRP letter to ARDSNet, signed by Kristina Borror and Michael Carome, addressed to Ronald Newbower and other investigators, July 3, 2003, in author's files.

Patz, A., L.E. Hoeck, and E. De La Cruz. "Studies on the Effect of High Oxygen Administration in Retrolental Fibroplasia. I. Nursery Observations." *American Journal of Ophthalmology* 35, no. 9 (Sep, 1952): 1248-1253. doi:0002-9394(52)91140-9 [pii].

Penman, D. T., J. C. Holland, G. F. Bahna, G. Morrow, A. H. Schmale, L. R. Derogatis, C. L. Carnrike Jr., and R. Cherry. "Informed Consent for Investigational Chemotherapy: Patients' and Physicians' Perceptions." *Journal of Clinical Oncology* 2, no. 7 (1984): 849-855.

Peppercorn, J. M., J. C. Weeks, E. F. Cook, and S. Joffe. "Comparison of Outcomes in Cancer Patients Treated within and Outside Clinical Trials: Conceptual Framework and Structured Review." *Lancet* 363, no. 9405 (Jan 24, 2004): 263-270.

Pinker, Steven. "The Moral Imperative for Bioethics." *Boston Globe* August 1, 2015. Accessed August 19, 2019, https://www.boston-globe.com/opinion/2015/07/31/the-moral-imperative-for-bio-ethics/JmEkoyzlTAu9oQV76JrK9N/story.html.

Pinn, Vivian. "From Exclusion to Inclusion—Participation in Biomedical Research and the Legacy of the Public Health Syphilis Study at Tuskegee." In Ralph V. Katz and Rueben C. Warren, eds, *The Search for the Legacy of the USPHS Syphilis Study at Tuskegee* (Lanham, MD: Lexington Books, 2011).

Prainsack, Barbara and Alena Buyx. *Solidarity: Reflections on an Emerging Concept in Bioethics*. London: Nuffield Council on Bioethics, 2011.

Pratt, Craig M. and Lemuel A. Moyé. "The Cardiac Arrhythmia Suppression Trial: Casting Suppression in a Different Light." *Circulation* 91, no. 1 (1995): 245-247.

Prentice, Ernest D., Sally L. Mann, and Bruce G. Gordon. "Administrative Reporting Structure for the Institutional Review Board." In *Institutional Review Board: Management and Function*, 2nd ed, ed. by Elizabeth A. Bankert and Robert J. Amdur. Burlington, MA: Jones & Bartlett Learning, 2006, 31-32.

Presidential Commission for the Study of Bioethical Issues. *Moral Science: Protecting Participants in Human Subjects Research*. Washington, D.C.: 2011.

Pronovost, Peter, Dale Needham, Sean Berenholtz, David Sinopoli, Haitao Chu, Sara Cosgrove, Bryan Sexton, Robert Hyzy, Robert Welsh, and Gary Roth. "An Intervention to Decrease Catheter-Related Bloodstream Infections in the ICU." *New England Journal of Medicine* 355, no. 26 (2006): 2725-2732.

Ramsey, Paul. *The Patient as Person: Explorations in Medical Ethics*. New Haven: Yale University Press, 1970.

Raper, Steven E., Narendra Chirmule, Frank S. Lee, Nelson A. Wivel, Adam Bagg, Guang-ping Gao, James M. Wilson, and Mark L. Batshaw. "Fatal Systemic Inflammatory Response Syndrome in a Ornithine Transcarbamylase Deficient Patient Following Adenoviral Gene Transfer." *Molecular Genetics and Metabolism* 80, no. 1-2 (2003): 148-158.

Reverby, Susan, ed. *Tuskegee's Truths: Rethinking the Tuskegee Syphilis Study*. Chapel Hill, NC: University of North Carolina Press, 2000.

Reverby, Susan. *Examining Tuskegee: The Infamous Syphilis Study and its Legacy*. Chapel Hill: University of North Carolina Press, 2009.

Rhee, Chanu, Travis M. Jones, Yasir Hamad, Anupam Pande, Jack Varon, Cara O'Brien, Deverick J. Anderson, David K. Warren, Raymund B. Dantes, and Lauren Epstein. "Prevalence, Underlying Causes, and Preventability of Sepsis-Associated Mortality in US Acute Care Hospitals." *JAMA Network Open* 2, no. 2 (2019): e187571.

Rhodes, Rosamond, Jody Azzouni, Stefan Bernard Baumrin, Keith Benkov, Martin J. Blaser, Barbara Brenner, Joseph W. Dauben et al. "*De minimis* risk: a proposal for a new category of research risk." *The American Journal of Bioethics* 11, no. 11 (2011): 1-7.

Rich, Wade D., Kathy J. Auten, Marie G. Gantz, Ellen C. Hale, Angelita M. Hensman, Nancy S. Newman, and Neil N. Finer. "Antenatal Consent in the SUPPORT Trial: Challenges, Costs, and Representative Enrollment." *Pediatrics (Evanston)* 126, no. 1 (Jul, 2010): e215-e221. doi:10.1542/peds.2009-3353.

Rich, Wade D., N. N. Finer, M. G. Gantz, N. S. Newman, A. M. Hensman, E. C. Hale, K. J. Auten, et al. "Enrollment of Extremely Low Birth Weight Infants in a Clinical Research Study may Not be Representative." *Pediatrics* 129, no. 3 (Mar, 2012): 480-484. doi:10.1542/peds.2011-2121.

Rohrig, Abie and Nir Eyal. "A New Day for Human Challenge Trials?" *Trends in Molecular Medicine* 28:7 (2022): 531-532.

Rothman, David J. *Strangers at the Bedside: A History of how Law and Bioethics Transformed Medical Decision Making*. New York, NY: Basic Books, 1991.

Savulescu, Julian. "Bioethics: Why Philosophy is Essential for Progress." *Journal of Medical Ethics* 41, no. 1 (2015): 28-33.

Scales Jr, Charles D., Alexandria C. Smith, Janet M. Hanley, Christopher S. Saigal, and the Urologic Diseases in America Project. "Prevalence of Kidney Stones in the United States." *European Urology* 62, no. 1 (2012): 160-165.

Schaeffer, M. H., D. S. Krantz, A. Wichman, H. Masur, E. Reed, and J. K. Vinicky. "The Impact of Disease Severity on the Informed Consent Process in Clinical Research." *American Journal of Medicine* 100, no. 3 (01/01, 1996): 261-268.

Schneider, Carl E. *The Censor's Hand: The Misregulation of Human-Subject Research.* Cambridge, MA: MIT Press, 2015.

Schrag, Zachary M. *Ethical Imperialism: Institutional Review Boards and the Social Sciences, 1965-2009.* Baltimore: Johns Hopkins University Press, 2010.

Schwetz, Bernard. "Interview with Tom Lamar Beauchamp, PhD." In *Oral History of the Belmont Report and the National Commission for the Protection of Human Subjects of Biomedical and Behavioral Research.* 2004. Accessed on October 9, 2022, https://www.-hhs.gov/ohrp/education-and-outreach/luminaries-lecture-series/belmont-report-25th-anniversary-interview-tbeacham/index.html.

Scott, James L., Gerald A. Belkin, Sydney M. Finegold, and John S. Lawrence. "Human Experimentation (Letter)." *New England Journal of Medicine* 275, no. 14 (1966): 790-791.

Select Committee on Intelligence, United States Senate. Project MKULTRA, the CIA's Program of Research in Behavioral Modification. Ninety-fifth Congress, First Session, August 3, 1977.

Senn, Nicholas. "A Plea For the International Study of Carcinoma." *Journal of the American Medical Association* XLVI, no. 17 (1906): 1254-1258. doi:10.1001/jama.1906.62510440008001a.

Sharav, Vera. "An Ethical Breakdown." Alliance for Human Research Protection. April 16, 2013, https://ahrp.org/an-ethical-breakdown, accessed February 13, 2021.

Sherwin, Susan. "Belmont Revisited through a Feminist Lens." In *Belmont Revisited: Ethical Principles for Research with Human Subjects,* eds. James F. Childress, Eric M. Meslin, and Harold T. Shapiro. Washington, DC: Georgetown University Press, 2005, 148-164.

Shoag, Jonathan E., Neal Patel, Lina Posada, Joshua A. Halpern, Talia Stark, Jim C. Hu, Brian H. Eisner, and Jonathan E. Shoag. "Kidney Stones and Risk of Narcotic Use." *The Journal of Urology* 202, no. 1 (2019): 114-118. doi:10.1097/JU.0000000000000197.

Sieber, Joan E. *Planning Ethically Responsible Research: A Guide for Students and Internal Review Boards*. Newbury Park; London; New Delhi: SAGE Publications, 1992.

Sieber, Joan E., Stuart Plattner, and Rubin Philip. "How (not) to regulate social and behavioral research." *Professional Ethics Report* 15(2), Spring 2002.

Silverman, William A. "A Cautionary Tale about Supplemental Oxygen: The Albatross of Neonatal Medicine." *Pediatrics* 113, no. 2 (Feb, 2004): 394-396. doi:10.1542/peds.113.2.394.

Silverman, William A. *Retrolental Fibroplasia: A Modern Parable*. New York: Grune & Stratton, 1980.

Skloot, Rebecca. *The Immortal Life of Henrietta Lacks*. New York: Crown Publishers, 2010.

Smekal, V., A. Irenberger, P. Struve, M. Wambacher, D. Krappinger, and F. S. Kralinger. "Elastic Stable Intramedullary Nailing Versus Nonoperative Treatment of Displaced Midshaft Clavicular Fractures—A Randomized, Controlled, Clinical Trial." *Journal of Orthopaedic Trauma* 23, no. 2 (Feb, 2009): 106-112. doi:10.1097/BOT.0b013e318190cf88.

Smith, Scott M., Martina Heer, Linda C. Shackelford, Jean D. Sibonga, Jordan Spatz, Robert A. Pietrzyk, Edgar K. Hudson, and Sara R. Zwart. "Bone Metabolism and Renal Stone Risk during International Space Station Missions." *Bone (New York, N.Y.); Bone* 81, (2015): 712-720. doi:10.1016/j.bone.2015.10.002.

Southam, Chester M., Alice E. Moore, and Cornelius P. Rhoads. "Homotransplantation of Human Cell Lines." *Science* 125, no. 3239 (1957): 158-160.

Speckman, Jeanne L., Margaret M. Byrne, Jason Gerson, Kenneth Getz, Gary Wangsmo, Carianne T. Muse, and Jeremy Sugarman. "Determining the costs of institutional review boards." *IRB: Ethics & Human Research* 29, no. 2 (2007): 7-13.

Stark, Ann R., Waldemar A. Carlo, Jon E. Tyson, Lu-Ann Papile, Linda L. Wright, Seetha Shankaran, Edward F. Donovan, William Oh,

Charles R. Bauer, and Shampa Saha. "Adverse Effects of Early Dexamethasone Treatment in Extremely-Low-Birth-Weight Infants." *New England Journal of Medicine* 344, no. 2 (2001): 95-101.

Stark, Laura Jeanine Morris. "IRBs and the problem of 'local precedents'." In I. Glenn Cohen and Holly Fernandez Lynch, eds, *Human Subjects Research Regulation: Perspectives on the Future.* Cambridge, MA: MIT Press, 2014.

Stark, Laura Jeanine Morris. "Morality in Science: How Research is Evaluated in the Age of Human Subjects Regulation." PhD dissertation, Princeton University, 2006.

Stark, Laura Jeanine Morris. *Behind Closed Doors: IRBs and the Making of Ethical Research.* Chicago; London: The University of Chicago Press, 2012.

Steinbrook, Robert. "How Best to Ventilate? Trial Design and Patient Safety in Studies of the Acute Respiratory Distress Syndrome." *New England Journal of Medicine* 348, no. 14 (2003): 1393-1401.

Steinbrook, Robert. "Protecting Research Subjects—The Crisis at Johns Hopkins." *The New England Journal of Medicine* 346, no. 9 (Feb 28, 2002): 716-720.

Stewart, David J. and Razelle Kurzrock. "Cancer: The Road to Amiens." *Journal of Clinical Oncology* 27, no. 3 (Jan 20, 2009): 328-333.

Stewart, David J., Simon N. Whitney, and Razelle Kurzrock. "Equipoise Lost: Ethics, Costs, and the Regulation of Cancer Clinical Research." *Journal of Clinical Oncology* 28, no. 17 (2010): 2925-2935.

Stokes, Joseph, J. Edward Berk, Leonard L. Malamut, Miles E. Drake, Jeremiah A. Barondess, Winslow J. Bashe, Irving J. Wolman, John D. Farquhar, B. Bevan, and RJ Drummond. "The Carrier State in Viral Hepatitis." *Journal of the American Medical Association* 154, no. 13 (1954): 1059-1065.

Stokes, Joseph, Jr. "Epidemiology of Viral Hepatitis A." *American Journal of Public Health and the Nation's Health* 43, no. 9 (Sep, 1953): 1097-1100. doi:10.2105/ajph.43.9.1097.

SUPPORT Study Group of the Eunice Kennedy Shriver NICHD Neonatal Research Network. "Early CPAP Versus Surfactant in Extremely Preterm Infants." *New England Journal of Medicine* 362, no. 21 (2010): 1970-1979.

SUPPORT Study Group of the Eunice Kennedy Shriver NICHD Neonatal Research Network. "Target Ranges of Oxygen Saturation in Extremely Preterm Infants." *New England Journal of Medicine* 362, no. 21 (2010): 1959-1969.

Sykes, Keith. "Fifty Years On." *History of Anesthesiology Proceedings* 34, (2004): 9-15.

Tavernise, Sabrina. "Premature Babies Study Raises Debate Over Risks and Ethical Consent." *New York Times,* September 7, 2015.

Terry, W., L. G. Olson, P. Ravenscroft, L. Wilss, and G. Boulton-Lewis. "Hospice Patients' Views on Research in Palliative Care." *Internal Medicine Journal* 36, no. 7 (Jul, 2006): 406-413.

The Church of Jesus Christ of Latter-Day Saints. "Risk Management Division Services." N.d. Accessed on November 15, 2022, https://www.churchofjesuschrist.org/study/manual/safety-health-and-environmental-manual/01/1-3-risk-management-division-services?lang=eng;  http://www.resourcepartnersonline.-org/.

Tobin, Martin J. "Culmination of an Era in Research on the Acute Respiratory Distress Syndrome." *New England Journal of Medicine* 342, no. 18 (2000): 1360-1361.

Tuskegee University. "About the USPHS Syphilis Study." N.d. Accessed November 3, 2022, https://www.tuskegee.edu/about-us/centers-of-excellence/bioethics-center/about-the-usphs-syphilis-study.

Tyson, Jon E. "Use of Unproven Therapies in Clinical Practice and Research: How can we Better Serve our Patients and their Families?" *Seminars in Perinatology* 19 no. 2 (1995) 98-111.

Tyson, Jon E., Michele Walsh, and Carl T. D'Angio. "Comparative Effectiveness Trials: Generic Misassumptions Underlying the SUPPORT Controversy." *Pediatrics* 134, no. 4 (2014): 651-654.

Tyson, Jon E., Nehal A. Parikh, John Langer, Charles Green, and Rosemary D. Higgins. "Intensive Care for Extreme Prematurity—Moving Beyond Gestational Age." *New England Journal of Medicine* 358, no. 16 (2008): 1672-1681.

U.S. Department of Health, Education, and Welfare. *The Institutional Guide to DHEW Policy on Protection of Human Subjects (the Yellow Book).* DHEW Publication No. (NIH) 72-102. Washington, DC: Government Printing Office, 1971.

van Beinum, Amanda, Nick Murphy, Charles Weijer, Vanessa Gruben, Aimee Sarti, Laura Hornby, Sonny Dhanani, and Jennifer Chandler. "Family Experiences with Non-Therapeutic Research on Dying Patients in the Intensive Care Unit." *Journal of Medical Ethics* 48, no. 11 (2022): 845-851.

van den Hoonaard, Will C. *The Seduction of Ethics: Transforming the Social Sciences.* Toronto: University of Toronto Press, 2011.

Vist, Gunn Elisabeth, Dianne Bryant, Lyndsay Somerville, Trevor Birminghem, and Andrew D. Oxman. "Outcomes of Patients Who Participate in Randomized Controlled Trials Compared to Similar Patients Receiving Similar Interventions Who do Not Participate." *Cochrane Library* 2010, no. 1 (Jul 16, 2008): MR000009. doi:10.1002/14651858.MR000009.pub4.

Wagner, Todd H., Christine Murray, Jacquelyn Goldberg, Jeanne M. Adler, and Jeffrey Abrams. "Costs and Benefits of the National Cancer Institute Central Institutional Review Board." *Journal of Clinical Oncology* 28, no. 4 (Feb 1, 2010): 662-666.

Ware, Lorraine B. and Michael A. Matthay. "The Acute Respiratory Distress Syndrome." *New England Journal of Medicine* 342, no. 18 (2000): 1334-1349.

Washington, Harriet A. *Medical Apartheid: The Dark History of Medical Experimentation on Black Americans from Colonial Times to the Present.* New York: Doubleday, 2006.

Weiss, Rick. "US Halts Human Research at Duke." *Washington Post* May 12, 1999, p. A1.

Whitney, Simon N. "A Fern in Amber: Risk Management in Research with Humans." *Risk Management* 17, no. 4 (2015): 226-239.

Whitney, Simon N. *Balanced Ethics Review: A Guide for Institutional Review Board Members.* Cham, Switzerland: Springer, 2016.

Whitney, Simon N. "Institutional Review Boards: A Flawed System of Risk Management." *Research Ethics* 12, no. 4 (2016): 182-200. doi: 10.1177/1747016116649993.

Whitney, Simon N. "The Python's Embrace: Clinical Research Regulation by Institutional Review Boards." *Pediatrics* 129, no. 3 (2012): 576-578.

Whitney, Simon N. and Carl E. Schneider. "Viewpoint: A Method to Estimate the Cost in Lives of Ethics Board Review of Biomedical

Research." *Journal of Internal Medicine* 269, no. 4 (Apr, 2011): 396-402.

Whitney, Simon N., Kirsten Alcser, Carl Schneider, Laurence B. McCullough, Amy L. McGuire, and Robert J. Volk. "Principal Investigator Views of the IRB System." *International Journal of Medical Sciences* 5, no. 2 (Apr 2, 2008): 68-72.

Wilfond, Benjamin S., David Magnus, Armand H. Antommaria, Paul Appelbaum, Judy Aschner, Keith J. Barrington, Tom Beauchamp, Renee D. Boss, Wylie Burke, and Arthur L. Caplan. "The OHRP and SUPPORT." *New England Journal of Medicine* 368.25 (2013): e36.

Wolf, Susan M., Frances P. Lawrenz, Charles A. Nelson, Jeffrey P. Kahn, Mildred K. Cho, Ellen Wright Clayton, Joel G. Fletcher, Michael K. Georgieff, Dale Hammerschmidt, and Kathy Hudson. "Managing Incidental Findings in Human Subjects Research: Analysis and Recommendations." *Journal of Law, Medicine & Ethics* 36, no. 2 (2008): 219-248.

Wootton, Susan H., Patricia W. Evans, and Jon E. Tyson. "Unproven Therapies in Clinical Research and Practice: The Necessity to Change the Regulatory Paradigm." *Pediatrics* 132, no. 4 (2013): 599-601.

Zywicki, Todd J. "Institutional Review Boards as Academic Bureaucracies: An Economic and Experiential Analysis." *Northwestern University Law Review* 101, no. 2 (Feb 1, 2007): 861-865.

# Index

# About the Author

S IMON N. WHITNEY, MD, JD, opened his family practice in rural Washington state in 1982. In 1995, his career took a sharp turn when he enlisted at Stanford Law School. He wanted to study and write about medical ethics, and for this a medical degree alone might not be enough.

At Stanford, he earned a degree from the law school, and also did a fellowship at the Stanford Center for Biomedical Ethics. In order to understand the place of ethics in the world, he served on the university's Institutional Review Board, which is charged with reviewing proposed research to make sure it does not abuse human subjects.

In 1999, Dr. Whitney accepted a faculty position at Baylor College of Medicine in Houston, Texas, which he held until 2022. He had dual responsibilities at Baylor—to see patients in the family medicine clinic, and to do original work in medical ethics. As he helped the scientists who were his new colleagues, he discovered the many problems with IRB review, and he began publishing papers proposing reform.

Whitney's view that the IRB system needs a major overhaul was unpopular among ethicists, and he sometimes struggled to get his papers published. But scientists were grateful that someone understood their difficulties. At the end of a talk at the MD Anderson Cancer Center, when Whitney asked if there were any questions. Emil

Freireich, a pioneering oncologist who had made breakthrough discoveries in the treatment of childhood cancer, asked simply "How can I help?" Other scientists have been equally enthusiastic.

In 2012, Whitney published an essay pointing out that IRB review had led, unintentionally but predictably, to lopsided enrollment in a study of the use of oxygen in the neonatal intensive care unit, leading to treatment recommendations that would apply less well to disadvantaged children. The article persuaded many scientists, but ethicists reacted with anger (one wrote that Whitney was part of a "recent wave of reactionary attacks" that is "at best ahistorical and at worse blindly hysterical").

Over time, Whitney realized that the system would never be changed by academic debate. As a first step, in 2016 he published *Balanced Ethics Review* (Springer), a manual for IRB members that showed how their oversight could be less harmful to the work of scientists. Now, his new book—*From Oversight to Overkill*—takes the debate over the IRB system's suppression of research into the public arena. It will enable readers with an interest in how science works, and how government action can harm the public interest, to learn the whole story and decide for themselves what should be done next.

Dr. Whitney welcomes comments and questions from readers. Please email him at swhitney@bcm.edu. You can learn more about him and his work at https://www.drsimonwhitney.com.